BUYING AND SELLING LABORATORY INSTRUMENTS

BUYING AND SELLING LABORATORY INSTRUMENTS

A Practical Consulting Guide

MARVIN C. McMASTER

A JOHN WILEY & SONS, INC., PUBLICATION

Published by John Wiley & Sons, Inc., Hoboken, New Jersey.
Published simultaneously in Canada.

Library of Congress Cataloging-in-Publication Data:

McMaster, Marvin C.
Buying and selling laboratory instruments : a practical consulting guide / Marvin C. McMaster.
p. cm.
Includes index.
ISBN 978-0-470-40401-0 (cloth)
1. Scientific apparatus and instruments–Marketing. 2. Laboratories–Equipment and supplies–Marketing. I. Title.
Q185.M36 2010
681′.750688–dc22

2009045873

10 9 8 7 6 5 4 3 2 1

CONTENTS

PREFACE

Buying laboratory instruments is one of the most expensive undertakings of the research director. A wide variety of instruments are required to deal with research problems, and to be effective, they must fit the laboratory's needs. A poorly selected instrument can dramatically affect the results produced and indirectly can affect the research papers produced, the quality of training provided to the investigator's students, and the investigator's chances for career advancement. There are major problems in ensuring that the customer is buying the right instrument at the best price and that the customer will be able to get vendor service and support to keep the instrument up and running to produce results.

Many people mistakenly believe that the research director selects and buys the major instrumentation for the laboratory. The principal investigator may pay for the instrumentation, but almost always a senior technician, graduate student, or postdoctoral student does the actual buying. These people are the ones who will be using the instruments and are most familiar with the laboratory's requirements. They investigate the available models, acquire specifications and price quotations, compare the services offered and the reputations of various manufacturers, and recommend the equipment needed. In many cases, they write the funding proposals and bidding specifications. An exception to this scenario is the case of a new professor or a new laboratory investigator, who has just come from a training environment, is acquainted with the selection procedure, or has not acquired an experienced senior technician. This person may make his or her own buying decisions.

My laboratory instrument experience began in graduate school and in a variety of postdoctoral studies. I used many types of analytical and separation instruments to purify and identify the research mixtures I was studying. I helped select new instruments as needed. When I moved on to company research laboratories, pilot plants, and production facilities, I recommended new analytical systems that were needed for my work and wrote bidding specifications to purchase the instruments. As a professional sales and technical support representative for 25 years for major instrument companies, I helped potential customers create funding proposals and bidding specifications, and I helped them select and order the needed instruments.

I was involved in the installation when possible, trying to ensure that the instrument was actually used, and also supported customers in solving their research and cost-for-test analytical problems. When I sold to either a university facility, a contract laboratory, or a commercial laboratory, I first made a courtesy call on the principal investigator, who then referred me to the buyer in the laboratory. There the actual selling started, and it continued until price negotiations were complete and a commitment to buy was made.

But the sales representative works for the vendor and has primarily the vendor's welfare in mind. It is true that vendors are interested in establishing a long-term relationship with the customer, but generally, they are not as interested as the customer in creating an exact fit of the least expensive solution to the customer's problem. Many vendors feel that their commitment ends when the instrument is shipped and installed in the customer's facility. They do not have a long-term commitment to making sure that the customer is successful in using the instrument.

At the moment, I occupy a unique position. I am a consultant and technical writer with no commitment to any instrument vendor. I have a broad technical background in medicine, and in organic chemistry, analytical chemistry, and biochemistry, that allows me to better understand the problems being investigated and helps me to serve as a guide to the purchase of the correct equipment.

I have always sold as a consular salesman, a partner in ensuring the laboratory instrument's successes. This is not a typical sales relationship. I have observed, after sales were completed, many instrument disasters and research misfits, and I have seen ways that these system misfits might have been avoided.

In Section I of this book, I try to guide instrument buyers through the shoals of funding, purchasing, and acquiring best-fit instruments at the least expensive price. This section provides information on how to find vendors that support customers with both knowledgeable service and application support. It also offers guidance on how to adapt existing instruments to new applications, how to automate and integrate these instruments with other instruments in the laboratory, and what to do with them when they move beyond their useful life in the research laboratory.

In Section II of the book, I provide a guide to the sales process to make the purchaser aware of what is going on and how to determine if the sales representative is either trying to help select the correct instrument for the customer's needs or manipulating the customer into purchasing the more elaborate and expensive system the vendor is trying to unload. I believe in a *win/win* sales process that leaves both the customer and the vendor satisfied with the sale.

Throughout my sales career, I have told my customers that a salesman like me, with a Ph.D. and my technical background, was an expensive investment for an instrument company. To be cost effective, I neeeded to sell them four systems even if they were only buying a single detector at the moment.. To make this happen, my goal was to make them so successful that they would expand that detector into a full system, buy again and again from me, and send me other customers who would buy additional systems.

That is not a common sales attitude, but it is a successful technique for building loyal customer accounts. Many salespersons focus only on the bottom line of the immediate sale, and meeting their company's monthly sales quota is their most important sale criterion. When the sale is made and the instrument is paid for, they feel no further commitment to the customer until the customer is ready to purchase another instrument. To me, this is shortsighted and destructive to a long-term sales relationship. Every selling opportunity is a new event for these *I win/you lose* salespersons, often carrying a negative burden from the last sale and leading to the public's perception of salespersons as belonging in a professional category somewhere between those of prostitutes and actors—certainly not someone customers would want working with their laboratory personnel.

If you insist on becoming a salesperson, approach this career as a professional and do not assume, like many, that "Anybody can sell. You just need to go out and do it." Learn to serve, study the profession, and become a true win/win salesperson. The techniques presented in this book can serve as a starting point, but you will need to find books by successful salespersons, study them, apply the information and practice, practice, practice until you improve.

A true win-win salesperson is someone you turn to when you want to solve a laboratory problem. This person should be able to help you define your needs and fill them with the right instrument at a fair price. He or she should help you get the equipment up and running and be there as a resource that the laboratory can use again in the future. Such a salesperson is a rare and valuable resource, and should be sought out and encouraged by referring him or her to other investigators seeking to purchase equipment in your institution. Like a good wine, this type of salesperson gets better with age and frequent use.

PART ONE

PURCHASING LABORATORY INSTRUMENTS

1

SELECTING LABORATORY INSTRUMENTS

As a young salesman, I usually got involved in buying fairly late in the selection process. Most often I began with a sales call when a potential customer responded to an interest card in a technical magazine, asked our company for a pricing quotation, or received a referral by a colleague.

I have talked to new customers after their system was purchased and installed, asking them how they determined exactly what type of equipment they needed for their laboratory to allow me to get involved earlier in the selection process. The success of my business depends on getting involved in the selection process as *early* as possible.

I found that the need for a particular laboratory instrument usually grows out of the design of an analytical protocol. An apparatus is needed to assay the completeness of a reaction's conversion and separate the components produced, achieve a desired compound purification, or analyze and identify a compound. A particular instrument is selected by (1) reproducing equipment used in similar separations in the customer's laboratory, (2) selecting a system recommended in a technical article or in a piece of literature, or (3) buying one recommended by a colleague doing a similar investigation. Some institutions had an individual well versed in a variety of laboratory instruments who acted as an in-house expert consultant to facilitate equipment-buying decisions.

Buying and Selling Laboratory Instruments: A Practical Consulting Guide.
By Marvin C. McMaster

Customers generally know what the equipment needs to accomplish. What they often do not know is exactly what a particular system can do, how to use it, and how well it matches their requirements. Customers want the system to solve their research problems at a reasonable cost; they do not want the equipment to turn into a research project that will consume the laboratory's time and resources.

Sales literature and sale representatives can and should offer an education on the instrument's features and how they may benefit the customer's research. Someone will need to sort through all the features/benefits to see if any of them will be useful to the laboratory now or in the future. A research project's need for a laboratory instrument usually begins with someone reading an article in a technical journal, either during normal reading or in a library search. It may also be triggered by an equipment display, a presentation, or a poster session at a technical meeting. Technical articles usually list the equipment and conditions used for a separation or an analysis. Some laboratories will simply buy a duplicate of that equipment, but usually these articles are simply guides suggesting the type of equipment needed for the work. The laboratory specialist in charge of selecting or recommending equipment will contact major manufacturers for literature on the specific instrument of interest. This will provide prices, features, and specifications to guide the specialist in making buying decisions and bidding specifications. This type of contact will also probably trigger a number of sales calls from other company representatives offering to help link features to benefits for the research and to sell the customer their instruments.

I am often asked why a researcher will usually buy from a major manufacturer rather than from one of many other companies that sell similar but less expensive hardware. My answer in the past, when I worked for major manufacturers, was that these companies became major manufacturers because they made the investment in time and money to do research, provide quality control, and offer a more complete package of hardware, service, and support. Responsive, knowledgeable, preferably local warranty service representatives from vendors are very important in keeping instruments up and running and fixing them when things go wrong, as they often can. Vendor-provided technical support laboratories and training schools offered by the major manufacturers can educate the instrument users about your laboratory and help keep your instrument from becoming a research project. This is especially true when the laboratory is unfamiliar with the type of instrument and its capabilities.

Smaller vendors lack the investment money to make this kind of commitment to help the customer succeed. They offer the hardware and expect the customer to handle the rest After an instrument becomes more generic or a commodity type and third parties or an in-house service department can repair it, it becomes safer to consider smaller companies as suppliers. These companies make sales by offering lower prices, lowering manufacturing specifications and quality control standards, and making their profit on volume sales.

For the customer, getting expert advice as early in the sale as possible is important in ensuring a successful instrument purchase. An alternative to the literature search is to go to a colleague's laboratory, explain your research project, and ask what equipment this person would recommend and from what source. This approach has been so successful that some large companies support a technical guru. This individual is usually an early adopter of this type of equipment who has made a strong effort to develop considerable expertise in selecting and using the instruments for his or her investigations. The guru is usually a tinkerer who has done research, has optimized his or her equipment, and knows how to apply it in a number of applications. Gurus know the players in the field, both in manufacturing and in research environments, and can help you determine the best instrument for your research problem.

Identifying this guru is very important for purchasing success, both for the prospective customer and for the sales representative. Nothing makes a customer more comfortable with a buying decision than the approval of a local technical guru. Listen to his or her advice; it is usually the fastest way to figure out exactly what you really need.

If you don't have access to such an individual, you may be able to get good advice from an outside consultant, a local university guru, or a technical sales representative. Consultants will obviously cost more but will be more objective if their information is current. Local university gurus may only be able to offer you theoretical information on the instrument, and if they have laboratory experience, it may only involve older equipment. Sales representatives obviously have a commitment to their employer, but they usually have access to current equipment and training. If they have actual laboratory experience of their own, they may be able to understand the research you are doing and be willing to act as an unpaid consultant for applying the instrument to the research. Take time to find out the representative's background.

Things change fairly rapidly in the technical instrument field. A manufacturer who led the field in innovation five years ago may have run into hard times, especially in this age of corporate takeovers and buyouts. Talk to the people in your company or in nearby universities who have recently purchased this type of equipment about the type of service and support they have received recently from the manufacturer you are considering. Get referrals to other users from your guru, consultant, or sales representative and call them to ascertain their level of satisfaction.

Very few laboratories, except those in cost-per-test facilities, such as environmental laboratories, can dedicate a piece of equipment solely to the application for which it was purchased. Things change as your research advances. If you are working in a cost-per-test laboratory, you will want to buy the most rugged, least expensive, simplest-to-operate instrument, work it to death, and then replace it with a similar instrument. Most laboratories need more flexible and therefore more complex and expensive systems. These can be reconfigured and used for a variety of applications. They often can be used for methods development application scouting to develop new uses for your instrument. These do-it-all systems are the lead instrument in the manufacturer's catalog, but often they cost as much as two simpler instruments that could each be applied to different applications.

Only the person who understands the research laboratory's goals and budget can make an intelligent decision on which instrument type provides the laboratory with the most bang for its buck. Ten years from now, you probably will not remember the cost of the flexible instrument that could handle many jobs, but you will always remember the cost of the simpler one that could not do enough.

There is always a decision to make when buying analytical equipment: (1) buy everything in one box or (2) buy a modular system. The system-in-a-box is usually cheaper, has a smaller footprint on the laboratory bench, and often comes as a complete solution for your current laboratory problem. Modular systems sprawl all over the laboratory bench, but they offer flexibility for expansion, component upgrades, and reconfiguration. They can be organized in a component rack, but they will always appear more incomplete and usually be more expensive than a system-in-a-box.

1.1 MODULAR SYSTEMS

Research systems are almost always modular in nature. When you buy a research system, you know your immediate needs, but you must plan for changes of direction that may occur and for new projects. You buy the state-of-the-art system because of its cost and appearance, but you are aware that instruments evolve and improve almost daily, especially in detector and computer technology.

Using high-performance liquid chromatography (HPLC) as an example, a typical research system might use two pumps, a computer-based gradient controller, a manual injector, and a variable-wavelength ultraviolet (UV) detector with data acquisition and processing done by a board inside the system-controller computer. Another configuration may involve a single-pump system using a controller-operated solvent-switching valve to provide programmable on-line access to three or four solvents for solvent gradients and column washout. Most system-in-a-box HPLCs will use the latter type of pumping system.

As your research progresses, you may need to add a secondary detector such as a coronal-charged aerosol detector (CAD) to analyze for phospholipids at a sensitivity significantly below what can be determined at 209 nm on the UV detector. If you bought the all-in-one system, you may find that the only place you can put the new detector is on top of or beside the box, like a lean-to addition.

Evolution does not always eliminate the need for older systems. Two-pump systems generally provide more reproducible gradients than a switching-valve system. Once you have worked out the solvent composition for a couple of separations, you may separate the two-pump gradient into two parallel isocratic systems by adding another detector and a second injector or an autosampler to automate one of the separations. The variable UV system may not be sufficiently sensitive or provide a definitive identification, so you may choose to add an in-line mass spectrometric detector with its own computer-based controller and data processor. This will provide the needed definitive identification and extend the laboratory's capability.

All of this will involve an increase in price, laboratory bench confusion, and the need for expertise and training. It is usually easier to justify the purchase of additional equipment for an existing system based on changing research needs than a brand new system. System-organizing racks can be purchased as the system is extended to control the bench-littering problem. When stacking mixed liquid-handling and electronic components, it is important to remember to keep the liquid-handling components such as pumps, injectors, and detectors below the electronic components. Liquids tend to leak, and electronic components such as computers are not always sealed against such leaks.

1.2 SYSTEMS-IN-A-BOX

Buying a system-in-a-box works very well in a cost-per-test laboratory with well-developed methodology. These systems are often sold as a turnkey system, including proven methodology protocols and all the supplies needed to turn the system on and run the first analysis. They may be sold as a system for a specific analysis, such as a carbohydrate analyzer, complete with step-by-step analysis protocols and everything needed for the initial analysis. They include consumable components, such as HPLC columns, filters, and solvents, that have to be replaced periodically, but they are generally very cost effecive. Cost-per-test laboratories require systems with automated components such as autosamples and a robotic sample preparation system to increase sampling accuracy, allowing around-the-clock operation, and decrease the need for operator intervention.

The problem with this type of system occurs when service is required or changes need to be made. The component requiring service is usually not accessible without tearing everything out of the box. Reassembly after repair must be carefully checked to ensure a return to smooth operation. Systems-in-a-box usually share common power supplies and a computer as the controller; when something happens, everything in the box is affected and out of operation. The system-in-a-box is difficult or expensive to retrofit with the latest technology if a technology leap occurs, and it is often left behind unless it simply needs a software upgrade. New components such as new detectors may not fit in the box.

It is usually more profitable for a manufacturer to sell you a new box rather than upgrade the old one. You are left with an old box filled with old technology, and you will need an operator to run outdated technology. This happened often with gas chromatography/mass spectrometry (GC/MS) systems equipped with mini-computer controllers. Operators were expensive to train and maintain, and the laboratory filled up with archived data stored on a variety of old computers. Many environmental laboratories have a graveyard of archived obsolete data systems that might be called on sometime in the future to produce information for a legal defense.

Cost-per-test laboratories operating on a three-shift, 24-hour day will often dedicate a system-in-a-box for a particular analysis to prevent sample cross-contamination. They can quickly justify the system's cost based on the analysis fees produced and will buy a new system-in-a-box if they add a new analysis to their portfolio or if a system needs to be replaced. If they do their own methods development, they probably have a modular component system somewhere in the laboratory for the development work and for problem troubleshooting.

1.3 AUTOMATION

Automated components are expensive, and there is always the question of whether automating the system is worth the cost or whether the money could be better spent elsewhere. University laboratories usually prefer to use graduate students to automate their equipment. Cost-per-test laboratories prefer adding computerized hardware automation allowing them to run minimally attended equipment around the clock and recover the expense of automation out of the increased profitability. Most environmental laboratories estimate that they can recover the cost of any level of automation within three to six months.

Automation can be added at the front end of a system in sample preparation and sample handling, such as in-line sample extraction, filtration, chromatography stations, and autosamplers. Total system controllers can integrate run start signals and valve openings. Back-end automation usually involves decision-aiding software to provide detector signal peak identification, data processing, compound identification, and report generation. Next-higher-system integration occurs when you move beyond automation of a single system to controlling data production from multiple systems of the same type from a single computer. The final automation integration is the laboratory instrument management system (LIMS) computer that pulls together instrument data, wet chemistry results, and typed-in sample data to produce a completed sample report for customers and for archival storage.

Computerized robotic sample treatment systems are programmable stations that can dilute, filter, extract, run simple chromatography with cartridge columns, and inject sample solutions into chromatography systems for analysis. Autosamplers are programmable injectors that hold sample solutions in vials until they are needed for injection. These autosamplers can also be programmed to periodically inject a series of standards used for sample concentration quantitization and automatically start the chromatography system components. They benefit a laboratory by replacing much labor-intensive sample handling and avoid errors occurring during repetitive manual workups.

Simple data processing automation systems perform peak detection on the detector's signal, identify peaks based on retention times, and generate a peak identification table based on signal strength versus retention times. The next level of data automation involves doing quantization calculations by comparison to signals from concentration standards and generating calibration and sample concentration reports. At the final level of data processing, the system will perform definitive compound identification from mass spectral data and produce a compound quantization and identification report.

The rule of thumb for automation is that the more complex it is, the more it will cost and the more things that can go wrong. It will always require human inspection and possible intervention. Although it generally can run unattended, if an error occurs it can rapidly generate large quantities of useless data. With sufficient use of internal standards, blanks, and surrogate compounds, it is sometimes possible to adjust and rescue a portion of the erroneous data.

1.4 DATA ARCHIVAL AND RECOVERY

Data sets are of little use unless they can be retrieved when needed. Data reports can be stored in easily recovered word processing formats, but there are times when the raw data are required as proof of analysis or for repeating a calculation or recalculation based on new standards or assumptions.

Data are usually stored in proprietary meta-files that vary widely from manufacturer to manufacturer and sometimes among various generations of machines from the same manufacturer. Public analysis laboratories are called on occasionally to defend their data in court. University research data can sometimes be challenged in the literature and require a defense.

Over the years, translation software has been written to allow popular proprietary formats to be translated into modern data formats that allow recalculation. The majority of raw data produced will never be needed for defense, so it makes no sense to translate all the archived data of a firm. Many laboratories employ storage rooms for old data systems that allow retrieval of data from long-dead and discarded systems. This is usually successful until the storage hard drive dies or the previous operator departs.

There have been attempts over the years to build a common data format (CDF) in which all chromatography data could be generated or into which the data could be translated. This CDF format was created, but it never became popular because manufacturers wanted to maintain proprietary formats to block future system sales to their competitors. For marketing purposes, it is easier to control a laboratory's purchases if the new, competitive system does not speak the same language as the laboratory's older systems. It was hoped that the introduction of the data integration LIMS systems would end this practice, but LIMS primarily accumulates and processes word processing types of reports and information.

2

STEP-BY-STEP PURCHASING

The individuals and the locations used are fictitious and are meant only as a guide to the purchasing process.

As a first approach to understanding the purchasing process, we will use a case study. Dr. Henry Jones is a newly appointed associate professor in the Chemistry Department at a University in Western Indiana. As a new faculty member, he has been provided laboratory space and a department startup grant of $150,000 for the purchase of equipment and supplies and for initial salaries for technicians and students. He must immediately begin to write grant applications if he plans to continue his research studies. He will be transferring limited grant funding from his previous position in the Chemistry Department at another University in Indiana. He has a five-year grace period to convert his current position to a tenure-track professorship based on his publication production and academic achievement and recognition.

Dr. Jones' research interests are in peptide and protein purification, specifically in the area of peptide hormones controlling renal hypertension. Dr. Jones is working in conjunction with the Nephrology Department of the University Medical School. He believes he is on the track of a specific hormone found in rhesus monkey pituitary extract, and he needs to concentrate and purify it in order to sequence it if it proves to be a peptide or a small protein. He will need to acquire equipment for purification, extraction, and identification and training for his students in using the

Buying and Selling Laboratory Instruments: A Practical Consulting Guide.
By Marvin C. McMaster

equipment. At the moment, he has only a former graduate student who is transferring with him to do a postdoctoral study at the University. The university has promised him an experienced technician who has been working with another professor, who failed to make tenure and moved to another university.

Dr. Jones' student, Dr. Tom Alonzo, is scheduled to go to a large technical meeting next month to attend seminars on protein and peptide purification and characterization. He also will be viewing posture sessions on the same subjects and visiting equipment booths in the exhibition hall in order to talk to vendor representatives and acquire literature on the potential types of equipment they will need. With extremely limited funds, gathering equipment pricing information is a critical part of his assignment. In the meantime, Dr. Alonzo will be spending much of his time in the university's library acquiring current information on polypeptide purification and analysis. Dr. Jones has his own assignments, which involve writing for grant extensions on his National Institutes of Health (NIH) grants and on new funding proposals to the Kidney Foundation.

Jennifer Branson, the technician assigned by the Chemistry Department, turns out to be a great asset. She has a BS in chemistry and a background in open column purification of proteins. She had worked on a project of purifying therapeutic quantities of angiotensin and bradykinase from uterine homogenates and is just recovering from walking pneumonia acquired from warking into and out of walk-in cold rooms. She knows how to make up buffer solutions and has some familiarity with instruments, but she is looking for work that will keep her out of the freezer. If Dr. Jones can attract a couple of serious second- or third-year chemistry students, he should have the core of his research group put together.

Tom Alonzo's library research has produced a number of recent research papers demonstrating how to purify a broad variety of peptides and proteins using high performance liquid chromatography HPLC and how to separate and analyze them using two-dimensional electrophoresis (2-D EP) and HPLC. He also has produced a list of the major suppliers of this equipment from the papers he has read. Dr. Jones talks to other researchers in the department's laboratories about the needed equipment and the companies that supply it. He receives glowing recommendations on some of the companies and their local sales representatives, as well as specific warnings from three different professors about one company to avoid at all costs. He asks his colleagues about the service and support they have received from each of the vendors that they use. Dr. Jones then calls his colleagues at the Medical School and talks to them about his findings concerning the needed equipment and the vendors that supply it.

Based on the input he has received, he calls the instrument manufacturers that Dr. Alonzo has mentioned and requests price quotations and literature from three HPLC suppliers and two suppliers of 2-D EP apparatus. Based on the ballpark verbal price quotes that he obtained on the telephone, it is obvious that the HPLC will make the biggest dent in his research funds and that the 2-D EP has the widest range of prices. Fortunately, the university has a core facility to provide fee-based peptide sequencing and molecular weight determination, so he will not have to make those major instrument purchases. The department also provides access to ultracentrifuge and scintillation banks if required. The spectrophotometers, clinical centrifuges, and pipetters that he brought with him should supplement the glassware that can be checked out from the department stores to complete his laboratory requirements for the moment.

The next step is to submit bids for the equipment that the laboratory needs. Dr. Jones sits down with Tom Alonzo and Mrs. Branson, who will be running the equipment. They review the literature from the instrument companies to sort out the critical specifications in the brochures that will affect their research, which they will need to include in the bidding specifications to ensure that they get only the equipment they want.

While they are going over the material, Dr. Jones receives a telephone call from a sales representative with one of the companies selling HPLC equipment and sets up an appointment to talk to him that afternoon, deciding that the salesman might be able to help them sort through the needed specifications. Dr. Jones meets with the representative and calls Tom Alonzo to his office, telling him to bring the HPLC brochures with him.

After introductions are made, the sales representative, Mitch Baker, asks Tom exactly what they are planning to do and how much money they have for the project. Tom explains that they are working on a limited startup budget and will need both a gradient HPLC and 2-D EP apparatus to do peptide and protein identification and purification. The sales representative asks for more details on the source and separation of the proteins and peptides. Tom spends about 30 minutes describing their research since the salesman's questions show that he understands the technology. Baker reviews the brochures and goes over the specifications, highlighting the critical ones for his HPLC system and for the 2-D EP. He compares the various pieces of equipment in the brochures, points out the features that will be important and the benefits they provide, and identifies the equipment that will be excessive for their current project and where it might be useful to the laboratory in the future.

Using the form that Dr. Jones obtained from the university's Purchasing Department, Mitch Baker and Dr. Alonzo work out the wording of the bid requirements and then add a list of vendors for the HPLC and 2-D EP systems. When Dr. Jones returns, Dr. Alonzo shows him what they have come up with. Dr. Jones initials the bid and urges Tom to have purchasing expedite it as quickly as possible. Tom makes a copy of the bidding requests for Mitch Baker to send to his Order Entry Department so that they have a heads-up on the bid's arrival.

Two weeks later, Dr. Jones receives the bid responses from Purchasing and reviews them with Dr. Alonzo. There is only one bid on the 2-D EP system and it closely matches their bidding specifications, so it is accepted. There are three responses to the HPLC system bid. One is a bare-bones, bottom-line system that fails to meet many of their requirements, especially for installation and training. The second bid is for a very expensive system with a diode-array UV scanning detector that was not in the bidding specifications; this bid is rejected because of its high price. The third bid is from Mitch Baker's company and exactly fits the bidding specifications. Dr. Jones accepts this bid and asks Tom Alonzo to write a justification letter rejecting the expensive bid and the insufficient system that will be sent back to Purchasing under Dr. Jones' signatures with the bidding package.

Tom calls Mitch Baker to let him know the decision, and asks him if he can expedite the bid through his company when Purchasing releases it. Mitch calls the head purchasing agent two days later to ask about the bid and to find out if he can mail the purchase order to his company. He stops by Purchasing the next day with an addressed, prepaid express mail envelope, receives the bid, and sends it to his company by overnight mail. Mitch then calls the company, books the instrument, gets it into the production queue, and alerts Drs. Jones and Alonzo to the arrival of the purchase order in the next day's mail. He gets a tentative ship date from Manufacturing and leaves the information in his field service representative's voice mail so that he can call Dr. Alonzo to set a date for installation. Then he calls Tom Alonzo to give him an approximate ship date and installation schedule and reminds Tom to inspect the shipping boxes as soon as they arrive for damage. He gives Tom the telephone number for an insurance claim in case it is needed.

About two weeks later, Dr. Alonzo walks into the laboratory to find a large pile of shipping boxes in front of the empty laboratory bench. He checks them for damage; except for one slightly crumpled corner, everything looks fine. He calls the number Mitch Baker gave him for booking the installation and then calls Mitch's voice mail to let him know that the system has arrived. The service man, Harold Wheeler, calls back and sets up the installation the next morning. Mitch calls back and makes an appointment for later that afternoon.

The salesman shows up with a laboratory-warming present: a surge protector for the new HPLC system. Tom mentions the damaged corner on the detector box, and Mitch asks for a box cutter and opens the top flaps on the box. He shows that the corner and the packing support corner on the inside are partially collapsed, but the instrument looks all right. He tells Tom to mention it to the service representative when he gets everything unpacked and set up so that he can check out the detector's response. Mitch hands Alonzo three sheets in protective plastic sleeves: a protocol for making and filtering solvents and samples, one for a simple isocratic sample run, and one for a simple gradient run. He tells Tom what standards, solvents, filters, and filtration apparatus he will need to order and gives him a couple of catalogs that offer the material he needs. He tells Tom that he will be out of town for a couple of days, but he will stop by the day after the installation to help Tom and Jennifer make the first run.

Setting up the 2-D EP is a different matter. Tom inspects the boxes for damage; they seem to be all right. No vendor's representative shows up, so Tom and Jennifer are left alone to pull out the pieces and hook them together. Tom had used a different EP apparatus at his previous school but had never put one together, so he knows that working with the high voltages involved can be tricky and potentially dangerous. He and Jennifer search the box, but all they find is a folded sheet that, when fully open, show step-by-step instructions on how to connect the pieces and cables. The only other instruction is a warning in three languages about the dangers of using high-voltage equipment. Jennifer finally notices a small 20-page read-me-first manual in the cable bag.

On the back of the manual is a telephone number for the company and the company's web site address. After considerable digging around on the site, Tom finds a 60-page operational manual for their particular model and downloads it to the department's laser printer. It provides a list of needed supplies, instructions on how to prepare working solvents, and advises the use of a three-prong surge protector for grounding and protection; none is supplied in any of the instrument boxes. With this setup phase behind him, Tom searches the HPLC manufacturer's web site and finds a similar, but much more generic, operational manual for their HPLC systems. This manual burns 180 pages on the laser printer. Tom obtains a three-hole punch and a couple of large three-hole binders, returns to the laboratory, and gives Jennifer a project for the rest of the afternoon.

Harold, the service man, shows up early the next morning. He opens all of the boxes, pulls out the cable bags, and then takes out the shipping invoice sheets, which he staples together and uses to check off each module, cable, and manual in each box as it comes out. He sprays out the box flaps, turn the boxes out on their open tops, slips the boxes up off the instruments, pulls off the packing corners and the shipping bags, and then lifts up each instrument and places it on the laboratory bench. He checks to make sure that fuses are in the back, sets the voltage for USA (110–120V), and plugs the power cable into the back. He arranges the modules by stacking the pumps, mounting the electronic controller and the detector above the liquid modules. He shows Jennifer and Tom how to make compression fittings on the tubing to connect the pumps, injector, column, and detector. He connects and runs signal cables from the controller and detector to individual modules and the data card in the back of the data acquisition computer. Finally, he connects the power cables from all the modules to the surge protector, turns on power switches for each module, and fires up the computer. Everything turns on and the appropriate gauges, dials, and lights come on, even on the suspect detector.

From his bag, Harold pulls out a coil of tubing with a fitting on it that he calls a *column blank*, removes the column, and asks Jennifer for a bottle of HPLC methanol. He puts the solvent line sinkers in the bottle, turns on the pump flow from the controller, and washes out the pumps, injector, column bridge, and detector flow cell with methanol into a beaker. He zeros the detector signal and turns on the computer to get a baseline reading. He then takes from his bag a bottle with a mixture of standards, holds the blunt-tipped injector supplied with the injector, and shoots a 20 μL sample of standard solution into the injector. Immediately, he sees a peak rise from the baseline on the computer screen. He says that everything is working and looks fine, and asks Tom to sign off on his installation report.

The final act of the purchase is carried out the next day. Mitch, the salesman, shows up, guides Jennifer through the process of making up a liter of 80% methanol/HPLC water, and filters it with the laboratory's vacuum filtration apparatus. He has Jennifer flush out the pumps through the injector into a beaker, make a new tubing connector from the injector to the column, flush the column outlet end into the beaker, and then connect the tubing from the column end to the detector. He shows her how to fill the syringe with 20 μLs of standard and get rid of bubbles by flicking the barrel of the syringe. Then he asks her to shoot the sample into the injector. Turning the injector starts data acquisition in the computer, and about five minutes later, the first of four peaks appears. Mitch asks her to enter the date, her name, sample ID, flow rate, column type, and detector settings into the software and then has her print out a chromatographic report. He congratulates her on becoming a chromatographer, buys her a Coke, and helps her set up her registration for the HPLC school at the company's headquarters. She and Tom Alonzo will both be going for a week-long hands-on training session.

Is this a typical purchase case study? It was when I was a salesman. I sold multiple instruments in the same account by making the customer successful. Other instrument purchases from our competitors that I saw were more like the purchase of the 2-D EP system described above. The instrument was delivered and showed up on the customer's desktop, and the customer had the the fun of making it work, if it did work!

Which purchasing experience would you prefer? You can understand the importance of doing the homework to find the equipment, the vendor, and the salesperson you need. This can free you to spend your time generating research results, publishing papers, and producing well-trained graduate students.

3

ANALYTICAL INSTRUMENT SPECIFICATIONS

Simple laboratory instruments such as water baths, magnetic stirrers, and melting point apparatus are commodity items and are easy to buy. You match the range of operation specifications to your needs and when you find a match, in a catalog, in a brochure, or on-line, you place an order and the instrument should fit the desired application. Your examination of product specifications must be more careful as an instrument's complexity increases. Specifications supplied by a manufacturer increase with increased complexity, but not all benefits offered by a particular specification are important to your application of the instrument.

Specifications are like rabbits; they tend to proliferate. People who develop these lists seem to be like the professor who graded papers by throwing them down the stairs outside his office. The heavier they were, the farther down they went, and the heaviest ones got the best grade. The price of any instrument seems to be linked to the number of features it has and the accompanying benefits that a company can offer, which leads to feature and specification glut in brochures and in manufacturers' literature. Your job in examining equipment for purchase is to wade through the specifications and decide which ones are important to the success of your project. Many critical but intangible features, such as on-site service, warranties, application support, and training, are offered only in passing in the literature and are very difficult to evaluate for quality and cost when

Buying and Selling Laboratory Instruments: A Practical Consulting Guide.
By Marvin C. McMaster

making your decision. The only help you will find in evaluating these items is in referrals from current users or from your company's gurus, whom you hope will offer current information. To aid in wading through specifications lists, I have provided in Appendix B, Table 1, a list of what I consider to be critical specifications for a variety of instrument types.

3.1 DEDICATED PACKAGES VERSUS COMPONENT SYSTEMS

Manufacturers will often offer solution systems in a single package, complete with software and application protocols, at prices considerably lower than those of modular systems, especially when the technology addressed by the instrument becomes reasonably mature. These systems are very attractive to the budget-strapped laboratory, but they suffer from a lack of flexibility and are often difficult to expand or update as technology advances. They tend to be a works-in-a-drawer design, and the newest equipment may not fit in the drawer. Critical specifications for any large system, whether dedicated or component, are size and weight. The system's footprint on the laboratory bench will always impact other uses of that space by the analysis group.

Systems are also designed in a very compact form to save bench space, which often makes them very difficult to service. One very early computerized, dedicated HPLC system had its UV detector buried in the lower back corner of the box. If the system was jarred, the detector slipped out of alignment and the system had to be practically gutted to repair or realign the detector.

Modular systems are usually more expensive and are easily expanded or modified as research changes. Components can be switched at will, and new components can be easily added. Modular systems offer the ultimate in flexibility for the research laboratory with changing application needs. Their downside is that they tend to sprawl across the laboratory bench unless accompanied by some type of organizer bench or rack. They also are usually a little rawboned, with tubes sticking out at random—not very attractive when the funding committee, alumni, or news reporters are touring the laboratory.

Interconnecting cables and a software-controlled computer can unite dedicated and modular systems containing components designed for external control. Either system's capability is dependent on the ability to upgrade the software as components change or are added. Dedicated systems usually link users to a single supplier that may offer state-of-the art advances. Modular systems usually offer more flexible or replaceable generic software to fit new capabilities, such as adding more sensitive detectors or gradient capability to an HPLC system.

Software evolves faster than any hardware component. Hard-to-use or inflexible software can lock users into yesterday's technology as their competitors move into new, more productive areas.

3.2 CRITICAL FEATURES OF LABORATORY INSTRUMENTS

Critical features of each type of laboratory instrument are tied to the major function of the instrument. The most important feature of a *hot plate, water bath*, *incubator*, *rotary evaporator*, or *melting point apparatus* is their ability to generate and control heat and the maximum temperature the instrument can reach. Important related specifications might be its temperature range, heating rate, and temperature-holding precision. Secondary feature might have to do with capacity and size, which affect the amount or number of samples that can be handled or the amount of laboratory bench space that the instrument would occupy.

3.2.1 Universal Laboratory Equipment

Critical features of *pipetters, scales*, and *balances* are their handling capacity and the precision with which they can weigh, measure, or deliver samples. Those of *laboratory mixers* and *magnetic stirrers* are capacity, stirring speed range, and torque. *Fume hoods* feature storage volume and air-changing speed; specialty application hoods may also have safety features such as blast resistance. *Centrifuges* and ultracentrifuges feature sample-handling capacity, speed, and centrifugal force production; safety features such as rotor failure containment are also critical specifications. *Refrigeration equipment* features temperature control range, minimum temperature settings, and sample capacity.

3.2.2 Spectroscopy and Analyzer Instruments

Critical specifications for *UV, infrared*, and *nuclear magnetic resonance spectrometers* center on analysis range, measurement precision, minimum sample capacity, cutoff wavelengths, and sensitivity. Capacity specifications deal with sample cell design, sensitivity with minimum compound detectability, usable sample solvents, and available wavelength or excitation range. The lamp's lifetime and nature are additional secondary features. *Scintillation counters* feature sample capacity, counting precision, and shielding specifications. *Protein, peptide*, and *DNA sequencers* feature minimum sample capacity and maximum sample chain length; also important are single cleavage-step precision and cleavage-product identification precision specification.

3.2.3 Separation Systems

Open column chromatography systems specifications focus primarily on column volume and sample capacity unless they are pumped systems that feature pump flow rates, detector sensitivity, and attenuation range. *Sample preparation and extraction (SPE) cartridges* specifications are bonded-phase nature, packing material quantity, and sample capacity. *Thin-layer chromatography (TLC)* features support media, plate size, sample capacity, and plate media reproducibility specifications. *Gel electrophoresis* and *2-D sodium dodecyl sulfate–(SDS)-Page systems* feature voltage range, sample capacity, and tube or plate media reproducibility specifications.

3.2.4 Definitive Chromatography Systems

Critical specifications become more complex for more expensive hybrid chromatography systems with a variety of detectors, automation, and flow controllers. *Gas chromatography (GC)* and *gas chromatography/mass spectrometry (GC/MS) systems* feature oven temperature range, ramping rate, and ramping precision; cooling types and rates; carrier gas type, flow, and pressure control; detector(s) sensitivity, precision, attenuation, and wavelength; MS lens settings; (EM) potential; and mass range. *High-performance liquid chromatography (HPLC)* and *liquid chromatography/mass spectrometry (LC/MS) systems* specifications of importance are pumping flow rates and pressure; gradient range and precision, injector sampling size; detector(s) sensitivity, precision, attenuation, and wavelength; MS lens setting; EM electromotive potential; and mass range. *Supercritical fluid chromatography (SFC)* and *SFC/MS systems* feature supercritical gas pressure, flow rate, and the same detector features as in GC/MS systems. *Capillary electrophoresis (CE)* and *CE/MS systems* feature voltage potential, flow rate, and the same detector features as in GC/MS or LC/MS. Additionally, systems with mass spectrometric detectors include features on minimum detector operating pressure, mass range, sensitivity, and precision.

3.2.5 Automation Accessories

Complex hybrid systems require computer-controlled automation for gradient generation, for external valve control, and for data acquisition and processing allowing unattended system operation. Critical features of *autosamplers* are injection size and delivery precision. Secondary features are injection repeating, needle wash control, standard interrupt cycle, definitive vial identification, and cooling range. *Chromatography system controllers* feature temperature control, multipump flow rate control, and external valve control specifications. *Data systems* and *laboratory information management systems (LIMS)* specifications are data acquisition rate, range, and precision; data integration parameters, such as noise rejection, peak detections, and slope; and peak retention time or peak mass identification. Report integration and outputting capability are secondary specifications. *Laboratory robotics systems* are automated sample handlers with specifications similar to those of autosamplers and autopipetters, as well as arm movement speed and precision parameters.

3.2.6 Mass Analyzer Selection

Selecting the correct mass analyzer is critical to your success when buying an analytical system with a mass spectrometric detector. When purchasing an analytical instrument containing a mass detector, it is important to balance the job requirements against the instrument cost to get the best bang for your laboratory dollar.

When these systems first came out this was an easy decision because the only analyzer available was the *quadrupole* analyzer. Over the years the analyzer has become smaller, with increased mass discrimination, sensitivity, and wider mass range, see Figure 3.1.

In the last few years, new mass analyzers have appeared that offer new choices for separating and analyzing samples. The *3-D ion trap* came out as a mass analyzer that could trap individual ions for fragmentation studies, sort of a poor man's tandem mass spectrometer. The *linear ion trap* recently appeared combining characteristics of the quadrupole and the ion trap but offering a much larger trapping volume leading to much higher ion sensitivities for trace substance analysis.

The *triple quadrupole* or tandem analyzers were developed from the quadupole to provide a technique for separating masses for further fragmentation and analysis of the fragment masses formed to aid structure studies. A variety of hybrid tandem combinations of ion trap, quadrupole, and linear ion trap have recently come on the market to further confuse the buying decision, all offering improved fragment mass analysis. These combination mass analyzers are available at greatly increased instrument costs.

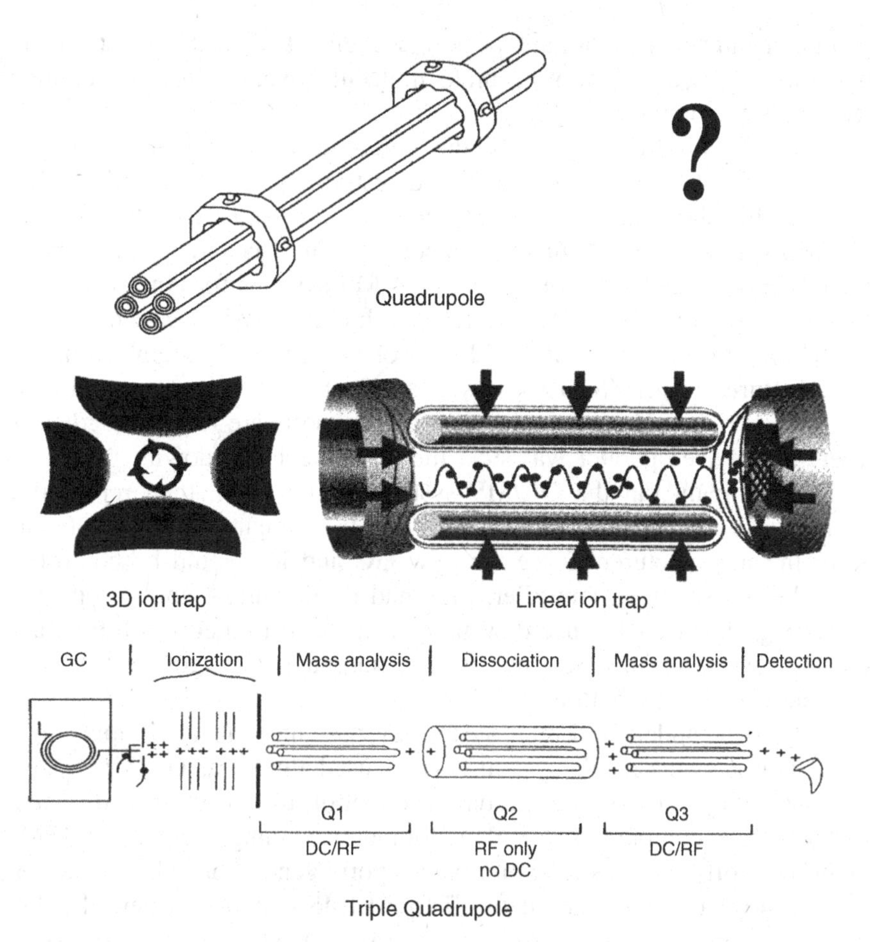

FIGURE 3.1 Mass Analyzer Selection

3.3 DEDICATED ANALYSIS FACILITIES

Core facilities in universities, large industrial research laboratories, and cost-per-test facilities have the same specification needs as the research laboratory. They use the same types of equipment, but their needs vary. They also acquire additional expensive, special-application equipment such as *molecular weight analysis systems, elemental composition systems*, and *accurate mass determination systems* that require specially trained personnel. They have the advantages of economy of scale and sampling volume in negotiating pricing, since they either purchase many machines of the same type or a variety of system types from a single purchaser. Again, all of these systems have critical specifications tied to their main function; for instance, accurate mass analysis specifications focus on the precision and accuracy of mass peak measurement, since these instruments generate two-decimal-point-accurate molecular weights used to confirm structures for publications.

I saw the difference between the purchase specifications for a typical research laboratory and a large Midwestern cost-per-test facility when I was selling laboratory instrumentation. The research laboratory had 13 GC/MS systems running three shifts per day. They had created the laboratory by buying old mini-computer-based MS systems for almost nothing, then replacing the control/data systems with modern Windows-based computer/software systems that could control two to four systems from any manufacturer's GC/MS. This was done so that any operator could run any system. The older mini-computer control software required dedicated operators; if the operator was sick, the instrument was down.

The company ran the GC/MS system until it died, then replaced it with a basic modern GC/MS system with no computer and no software from the manufacturer. Since the new GC and MS could be controlled with the old software/controller, they had no downtime and no operator retraining expense. Because they bought many instruments at a time and created their own specifications for the purchase, they got exactly what they needed at rock bottom prices.

They then modernized their system by building a local computer network that networked all systems. They used the local GC/MS system computer only for control, and data acquisition moved the raw data files off the local computer to another computer running the same GC/MS Windows software for integration and reports generation. Their next step was to move the integrated data off to a LIMS computer where the data could be combined with customer information, wet analysis, and other laboratory-generated information for creating customer reports.

4

FINDING THE BEST PRICE

Everyone would like to get the best deal possible when spending research grant money for very expensive laboratory equipment. Your results, your papers, and your career all depend on your instrument producing successful results.

There is always a fixed amount of money that a manufacturer must obtain to recover the cost of manufacturing and selling the equipment. These companies must make a profit; otherwise, they will not be in business the next time you need service, support, or a new instrument. Most of all, the manufacturer's equipment must be priced in line with that of major competitors. If the price seems unreasonably low, it is because the manufacturer is leaving something out of the hardware, service, and support package. In instrument purchases, as in life, there is no such thing as a free lunch.

As a general rule, the cost of manufacturing a product is 30% of the asking price. The balance of the price goes into sales margins or profits, administrative overhead, advertising, profit, warranties, and assistance options such as providing training and customer support laboratories.

Buying and Selling Laboratory Instruments: A Practical Consulting Guide.
By Marvin C. McMaster

4.1 PRICE QUOTATIONS

It is fairly easy to get the best instrument if best price is not an objective. You ask the manufacturers for pricing quotations. You then sort through the descriptions of features and specifications, hope you have found the ones that most closely fit your needs, and place an order. If you are using a bidding system, you determine the features and specifications that best fit your needs and then use these to write a lockout bid to ensure that you get the system you want.

Finding the best price for a particular instrument is not as easy. This will depend on your ability to negotiate, the amount of instrumentation you are buying from the same company, the amount of equipment you are buying or have bought from that company, the ability of your purchasing agent to negotiate discounts, and how close your sales representative may be to making his or her monthly or yearly quota.

4.2 GOVERNMENT SERVICE ADMINISTRATION (GSA) PRICING

I once asked an established customer how he was sure he was getting the best price. He said that he always checked with a friend who works in a laboratory owned by the federal government. His friend purchases all of his equipment with government funds off of the Government Service Administration (GSA) price list. The GSA is the largest purchaser of laboratory equipment in the United States. They negotiate all their prices and insist that they be given the very best price. A manufacturer who sells a particular piece of equipment below the price offered to the government can and will be removed from the GSA price list. According to industry rumors, companies have been fined seven-figure penalties for selling new instruments as demonstration equipment below the GSA price. They paid if they wanted all of their equipment placed back on the GSA price list.

The GSA price list is confidential and is to be used only by federal employees. But the government encourages cooperation between its facilities and university and private laboratories. If, in the process of this collaboration, colleagues happen to see a GSA price list, who is going to prosecute them?

4.3 INSTRUMENT SELECTION

We buy instruments to solve a research problem, to make a separation, or to provide information for an analytical application. Having gathered the information we need from manufacturers, the literature, or colleagues, it is time to fit that information to our particular application. If we are setting up to duplicate an analytical protocol for an environmental analysis, our job is fairly straightforward. We simply need to order the exact equipment specified in the Environmental Service Administration (ESA) protocol. Even this situation must be examined, however, since equipment evolves with time, as do protocols. There may be a better, cheaper system for getting the same result or a faster result while still adhering to the protocol. Information you obtain from a referral or the literature may be outdated. The research problem you are investigating may be completely different from or more demanding than the one handled by the company guru or the ESA protocol.

4.3.1 Fitting Your Needs and Budget

After you have gathered all the needed specifications, capabilities, and pricing information, you much match them to the requirements for your analysis and the budget for your instrument or your fund bidding. Your requirements need to cover your immediate needs and possible laboratory needs for this type of equipment for the next five years. You will also need to factor in the support and service capabilities of the company from which you are planning to buy.

If you are shopping for a UV spectrophotometer, knowing its application will determine its required features and eventual cost. If you plan to only do protein assay at 280 nm, you could select an inexpensive fixed-wavelength or a variable-wavelength instrument with the desired level of sensitivity. A few years ago, you would have selected a spectrophotometer with a sodium lamp and a 280 nm filter, but the lamp life for deuterium lamps used in variable-wavelength systems has increased dramatically. The capability to look at wavelengths ranging from 190 to 400 nm might be important for other applications. If you anticipate the need to look at colored samples, you might want to buy an instrument with an optional mercury lamp to extend your wavelength capability to 700 nm. Anticipating a use for GC/MS equipment, you might need a scanning UV spectrophotometer to find suitable wavelengths for target and secondary compounds to aid in compound identification. A spectrophotometer with nitrogen-purged optics would let you examine unstable compounds with absorbance below 190 nm. All of these extra features will come at a higher price than a basic spectrophotometer. Only you can judge whether their benefits to your laboratory's present and future applications are worth the difference in cost.

If the laboratory needs a separation system, the decision becomes more complex. Simple separations of a colored mixture can be made in a pipette or an open column filled with silica gel in a solvent, on a disposable TLC plate with compound recovery by scraping and extraction, or on an SPE cartridge attached to a syringe. If the compounds to be separated are colorless and similar in chemical nature, you may need a GC or HPLC system with high-performance columns and a detector capable of seeing the compounds to be separated.

The decision to use HPLC instead of GC is made because the samples to bc cxamined have a high melting point or are thermal labile, If the separation is well known, you may find a protocol in a technical journal or a manufacturer's brochure and buy a simple isocratic HPLC with a pump, injector, C_{18} column, variable UV detector, and integrator. If speed is important, you could buy a 3 μm column, but make sure that all solvents and reagents are purified through a 0.2 μm filter. If analysis speed is critical, you may decide to buy an ultra-fast HPLC, but your budget will take a big hit. If instead you require definitive identification of the separated materials, you might need an LC/MS system. The MS detector and evaporative interface are very expensive and require a trained operator to interpret the results, but they are the only instruments that will provide molecular weights and spectral confirmation of the compound's nature. For many complex separations and for use in methods development, you will need to purchase a gradient HPLC system, which may add more pumps, switching valves, a controller, and additional cost.

4.3.2 Consider Service and Support

Some laboratories have sufficient spare parts and trained service personnel to handle equipment emergencies, but most do not. Developing a new separation of very similar compounds is a research problem in itself, one that many laboratories would prefer to leave to separation specialists. Some manufacturers offer both services. Both of these needs must be factored in when considering buying equipment. How critical is it for the instrument to be back up and running in one to three days?

An extreme example of the service question arose with a customer I called on. He had a second laboratory set up in Antarctica and needed an HPLC system to analyze glacial meltwater samples. The turnaround time on spare parts was 14 months. I worked out a spare parts list for him, but to ensure that he would not lose a year's work, he purchased a backup isocratic HPLC system to take with him to ensure continued operation in addition to his spare parts purchases.

4.4 DEMONSTRATION EQUIPMENT DISCOUNTS

The 10% demonstration discount is the most commonly used form of discounting in laboratory instrument sales after the institutional bidding system.

I performed many instrument demonstrations to close sales, and I took care of my equipment. At the end of the demonstration, I washed out and stored my HPLC system's pumps and columns in 25% methanol/water. I tried not to run demonstration solvents containing either salt or buffer, and I sent my pump heads to the company for replacement after six months. Even then, there was a lot of wear and tear on an instrument carried around in the back of a station wagon strapped to a sales cart. I once received an integrator shipping carton shipped on an airliner after an equipment exhibition with forklift holes in the side of the box. The side of the integrator was bowed in, and the microprocessor boards on the inside were all snapped in half.

Most salespeople do not take good care of the company's equipment. I once saw an HPLC system that was accidentally dropped down the back staircase of a university department building strapped to an instrument cart. It ended up smashed, face down, on an asphalt driveway because the elevator was out. Pump heads and the injector were broken off, and the detector range selection buttons were jammed. I'm sure this system would make a nice addition to your laboratory.

If you are considering buying such a system, get its history. Look at the system in question. Talk to the sales representative who used it. Ask yourself how much you will save and how long you plan to use the system. Find out if it comes with a full new-instrument warranty. Never buy a demonstration instrument over six months to a year old that was only driven to church on Sundays by a little old lady.

4.5 DISCOUNTING IN KIND

A safer discount is the supplies discount or a discount on training and application development. Like car manufacturers, instrument companies have a higher profit markup on consumable supplies than they do on instruments. Often they will negotiate on supplies like solvents, columns, software, tubing, and fittings, especially if they manufacture them. Operator training is another useful service often thrown in to sweeten the deal. I used to offer a free two-day training and consulting session with each system sold. If enough systems were sold to a single account, I provided the training in the customer's facility. The first company I worked for provided a five-day hands-on training seminar for customers at the company's headquarters. The customers paid for their own travel, meals, and housing. It was good public relations for the company and tended to create lifetime customers for future purchases, and it is not surprising that the company had about an 80% market share in their instrument field.

Most companies will install a customer's system as part of the sale, but check to make sure that you will not need to negotiate this service. I found that an installed instrument is not successful until the customer makes the first run. An instrument that is not being used is a negative both for the customer and for the salesperson's chance to make another sale to the same account. I prepared a simple one page, step-by-step protocol sheet for a basic first run and walked the user through it. Often, I came back to see the sheet laminated in plastic and hung above the instrument for operator review and to aid in training other users.

A very few instrument companies maintain application laboratories that will develop new application protocols for a customer, sometimes for free but usually for a nominal fee. These protocols can save the customer research time and the need to purchase complex, expensive equipment for methods development, not to mention hiring a trained, experienced development chemist.

4.6 THE MODULAR TRAP

One of the major traps in buying modular instrument systems is to shop with many suppliers for the *best price* on each module. Some instruments work better together than others. It is often difficult to automate modules from a variety of manufacturers. If anything goes wrong with the system, mismatched modules often lead to a terminal outbreak of finger pointing. No one wants to take responsibility for the problem. The buck always seems to stop on your desk. At least when you buy all instruments from a single source, you know who is responsible for making them all work together.

4.7 BUYING USED EQUIPMENT

Again, let the buyer beware. Used equipment has only one saving grace—it is cheap. It is probably obsolete or discontinued equipment, which will make it difficult to find repair parts. If it is running at all, it is probably suboptimal; otherwise, the previous owner would still be using it. If it possesses computer software, this will usually be obsolete and probably very difficult to learn and use. Also, it may not run on your laboratory computer or have enough memory space on the build-in computer. That is the nature of old computers and software. Your time will be better spent writing grant proposals or otherwise searching for more money to buy a new system. The only used system worth the money is a free one.

4.8 NEW SYSTEM WARRANTIES

Warranties are written by lawyers to protect and limit manufacturers' liability. They are as good as the current reputation of the company behind them, and they are probably critical to your instrument success when Murphy's law intervenes.

Things will go wrong with even the best instruments, and usually the manufacturer's service department is your last, best hope. Before you call the doctor, it is important to isolate the problem as much as possible. I have found that 80% of all HPLC problems are column problems and 70% of those are due to bad water. Using fresh HPLC water, a correctly prepared column inlet line, and a new column can often make even the worst instrument problems disappear.

Service technicians come into your laboratory to fix instrument hardware problems, usually electronic problems. If the problem is in the application or in the chemistry, he or she will be in your laboratory for a long time with the meter running. A little time spent in thought can often save you a lot of money when it comes time to pay.

The best strategy to use is divide and conquer. Simplify the running system as much as possible. If you have a malfunctioning computer system, strip off accessories and applications until you have a minimal operating system. Plug it directly into the wall by removing the surge protector (it also goes bad). Reload the operating system and if everything works, start adding back the surge protector, applications you absolutely need, and then accessories until you find the one causing the problems. If the stripped computer with a new operating system does not work, it is time to tear it down or start looking at hardware components such as RAM memory and hard drives. Alternatively, get a new computer. How much is your time worth?

When working with HPLC systems, I found that the biggest problems were (1) solvents (especially water) and (2) bad columns. First, start with only the basic components needed to get a chromatogram. Make fresh solvents with HPLC water and no buffers or salts added. If you have one, use a fresh chromatography column and shoot a mixture of clean standards. If this system does not work, start checking components one at a time, from the data system toward the pumps. It is just like eating an elephant: start at one end or the other, one bite at a time.

Simple systematic diagnostics like this can save many hours of service time. I tried to get the companies to provide our service technicians with fresh columns and solvents as part of the service kit to be billed as part of the service call. I never succeeded, but, after all, the company gets paid for the service time that is billed on the meter.

5

GRANTS AND BIDDING

Competitive bidding is a process that allows states, universities, and companies to get the best price for a laboratory instrument by comparing offerings from many suppliers. Bidding contracts are prepared by purchasing agents working with researchers who need new equipment. These contracts describe exactly the equipment needed and when bids must be submitted. The contracts are submitted to a number of companies that supply this type of equipment, and open bids are posted in purchasing departments for inspection by prospective bidders. After the bidding deadline, bids are opened and examined by the submitting bidder to see if they meet the bidding requirements, a winning justified low bid is selected, and the bid is awarded to the winner. Vendors realize that the only way they can win a bidding contract is to offer their least expensive system that meets the bidding specifications and hope that no one else can meet their price. They offer the best discount that they can legally provide.

The assumption is made that all the systems being offered are similar enough that price alone can be used to choose among them. The wisdom of this assumption can be judge from a story concerning the space program. One of the Apollo astronauts was asked what he thought about as he lay on his couch waiting to be shot into space. He said all that he could think about were the million and one parts that made up the rocket beneath

Buying and Selling Laboratory Instruments: A Practical Consulting Guide.
By Marvin C. McMaster

him—all of them bought at the lowest price. When everything is on the line, you want the highest quality, not the cheapest product you can buy.

Prospective buyers must ensure that the bidding specifications they use when putting out a purchase on bid are written in such a way that they can be confident that the results compare apples with apples and oranges with oranges. Otherwise, they will end up buying the cheapest piece of fruit that anyone can come up with or having to rebid with tighter bid specifications. It is usually better to do it right the first time.

5.1 LOGICAL BIDDING SPECIFICATIONS

Logical bidding specifications come out of your research on specifications described in Chapter 3. You must sort through all of the detail concerning the features and benefits offered in the literature, brochures, and sales information presented by the manufacturers to find the ones that are important to you and your research, both present and future. Once these features have been boiled down to the essence of your needs, you write lockout specifications that the equipment you seek *must* have. When bids come in, you must inspect them to ensure that they do indeed include all the specification you defined before you choose the one with the best price. If the submitted bids do not have the correct specification, this can be used as a reason to reject the bid.

Lockout specifications can often be used to buy only a single system from a single manufacturer. This can be a dangerous procedure because it tends to eliminate newer technology with better features that may enhance your research results. Thing change with time, and sometimes change is good. The trick, of course, is to write specifications that are tight enough to provide the features you need without eliminating systems that you are unaware of with better features. If at all possible, include in your bidding process intangibles such as service contracts, application support and training, and warranties.

5.2 DEALING WITH PURCHASING AGENTS

Purchasing agents support and execute the bidding process and are supposed to be on your side in the negotiations to get the best products for your money. They do not have your technical expertise, and they will need your help in evaluating the bids when they come in. Their tendency is to see only the bottom line. If you let them make the final judgment, you will receive the least expensive item that has been bid, appropriate or not.

Purchasing agents are human, not computer-driven. They have favorite vendors, and they can sometimes get favored treatment and quantity discounting from vendors who supply large amounts of equipment to your account. This can be useful to you in getting a good price.

However, it is important to realize that there is no such thing as a free lunch. There is a limit to ethical legal discounting. One of the oldest forms of discounting was the baker's dozen. Often a baker would throw in a 13th doughnut to sweeten the deal for a regular customer. The customer would get 13 doughnuts for the price of 12, or about an 8% discount. The typical demonstration discount is 10%. If the discount is made in kind with higher-margin accessories or consumables, it may be slightly more than 10% but not much.

When submitting bids, always check your posted bids to make sure that all of the companies you are interested in are on the bid submittal list. Bids are not truly competitive if some of the interested vendors are left off the bid list.

5.3 USING GSA PRICING

As mentioned before, the U.S. government is often the largest-volume customer for any vendor. The government negotiates a single GSA price for all of the equipment their laboratories purchase and circulates GSA price lists to all of their laboratories. Manufacturers gain immediate access to all government instrument buyers by qualifying for inclusion on the GSA price lists. It is possible for a government laboratory to buy instruments that are not on the GSA price lists, but considerable justification must be created to do so. The process is generally much easier and more rapid if laboratories conform to GSA purchasing and buy off the list.

The GSA states that their price must be the best price that the vendor offers to anyone. If new instruments are sold at a lower price, the vendor can be removed from GSA bidding and will have to pay a significant fine to be reinstated. If the price you are offered is lower than the GSA price, an excellent reason must be given. Finding the GSA price for the item you are seeking by obtaining a copy of the GSA price list is always a good way to find the very best discount for the instrument you are seeking. You may not receive that price, since you do not buy in the volume that the government buys, but at least you will be able to evaluate your purchase to see if it is a good deal. The government does not encourage circulation of their price lists to nongovernment laboratories, but their availability does create cooperation among government and civilian researchers. Find a friend at a local meeting, conference, or users group who works for a federal laboratory or contract facility and ask him or her for help.

Other countries have different bidding systems and purchasing agreements. The French bidding system is said to welcome bidding, but it does not settle for the best price. The French average the bids and take the price closest to the average. They feel that this practice provides a representative price while eliminating shoddy or overpriced merchandise. GSA discounting levels may not bind foreign suppliers, but they have shipping costs to factor into their pricing and service and support considerations must be factored into their bottom-line price, so they generally are not able to compete with this discounting level.

5.4 QUANTITY DISCOUNTING

The more equipment you can buy in a single bid, the better the discount you can often negotiate. Bundling similar bids from a number of researchers can be advantageous. But often your requirements and bidding specifications will not fit your neighbor's needs. You must remember that a camel is a horse designed by a committee. If a camel can meet your laboratory's needs, you can probably get a good deal on it.

You may be able to work out a better discount by purchasing multiple instruments at one time if a single manufacturer produces or sells a number of different laboratory instruments that you can use in your research and if your budget can tolerate that large a cash outflow. A forward commitment for future purchases from the same manufacturer can sometimes gain you favored consumer discounts, but realize that people and situations within companies change over time. Few manufacturers are willing to put such relationships on paper.

A number of manufacturers are willing to write discounted forward-commitment agreements for consumable items such as separation columns, fittings, syringes, separation cartridges, filters, and solvents. Regular delivery of these items can be arranged, and these deliveries can be adjusted as requirements change, usually with a provision that an annual purchase level must be exceeded or part of the discount will be back-billed. These contracts benefit the manufacturer by providing a known revenue stream, and they benefit the customer by reducing costs and paperwork and by providing supplies as needed without reordering. Generally, these agreements are most beneficial to very large accounts or to laboratories providing cost-per-test analysis. Manufacturers like these commitments since the forward commitment locks the buyer's account into that manufacturer and eliminates any competition for those products. They are advantageous to the buyer because he or she does not have to rebid these items and avoids additional paperwork. They can become a problem if a major technological change occurs, so make sure to build in cancellation or volume adjustment arrangements in writing the agreement.

6

INSTRUMENT VENDOR SUPPORT

A successful customer is the goal of most professional sales representatives. A customer who is happily using the representative's equipment will refer him or her to other users in the account, and will mention the company in papers and at technical meetings. When the customer needs another instrument, he or she will think of the salesperson's company first.

Instrument success comes from three main factors, as shown in Figure 6.1. The instrument fits the needs of the application and the protocol that the customer is running (Hardware). The instrument is correctly set up and in working condition (Service). The customer's instrument operator knows how to use the instrument and how to get the best out of it in running the application (Support).

As computers and microprocessors have entered the field of instrument analysis, a second parallel success triangle has emerged. The customer must have the correct program (Software) to run the instrument; the software must be working correctly to acquire, store, and process the customer's data; and the customer must be trained to use the software to get the needed results.

Buying and Selling Laboratory Instruments: A Practical Consulting Guide.
By Marvin C. McMaster

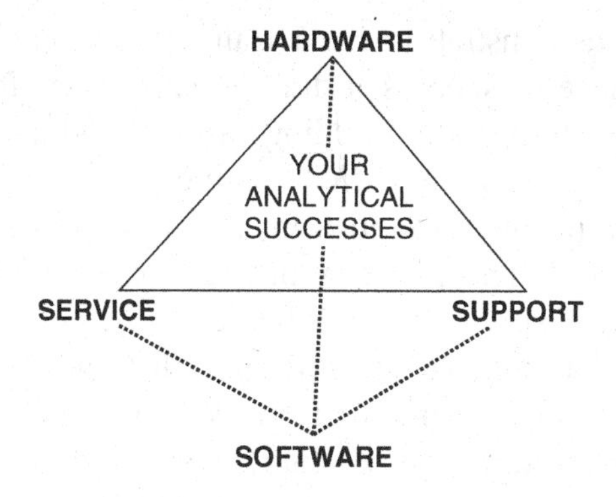

FIGURE 6.1 Instrument Success Pyramid

6.1 IN-HOUSE DEMONSTRATIONS AND SEMINARS

The first contact of a prospective buyer is usually with either the salesperson or a technical specialist in a support mode. The buyer needs to be convinced that the instrument he or she is considering will do the job. Journal articles and technical meetings help convince the buyer that your type of instrument can solve the problem, but with new instrument technology, the buyer must be further convinced.

An in-house slide-based seminar can often demonstrate the breadth of the instrument's analytical possibilities with real-life results. If the examples are close enough to the application needs of the buyer, they may be sufficiently convincing to lead to an order or to a desire to prepare bidding specifications. With established technology, this often can be supplemented with user referrals and testimonials.

But with newer instruments, nothing is more convincing for the prospect than to see compounds separated on the vendor's instrument in the buyer's laboratory. The instrument demonstration is a powerful convincer, but it can be filled with problems that can block the buying process. Murphy's law rides on every demonstration cart.

The dropoff instrument demonstration is the most dangerous of all demonstrations and the one most likely to fail. The customer does not own the instrument, does not know how to use it, and is often reluctant to turn it on for fear of breaking it. The customer does not know how to adapt the instrument to solve his or her problem unless he or she is already using similar equipment. Analytical equipment is not usually designed for transport, and components may have been knocked out of alignment. The dropoff demonstration works best with very simple instruments that have a limited number of features and components that can go wrong.

An instrument demonstration by a trained technical specialist has a slightly better chance of success. I did demonstrations for over 20 years from an instrument strapped to a folding cart that slid into the back of my station wagon and locked in place. Most demonstrations worked very well because I was familiar with the instrument and kept it serviced between demonstrations. But things still go wrong: cords get left behind, solvents go bad, and buyers' expectations can be very unrealistic.

It is important that both the support specialist and the buyer mutually agree upon the goals of the demonstration. When I was new to the game, I did a demonstration in which the customer wanted to see the separation of four different types of compounds, one of them at the advertised sensitivity limits of the instrument and another that required considerable sample preparation to be run. A one-day demonstration turned into a two-day nightmare, only a few of the customer's goals were reached, and to my knowledge, no instrument was ever purchased by the customer from any vendor. To make matters worse, the customer's laboratory was located at the far end of my sales territory.

Demonstrations should be done on a single class of compounds, at a reasonable sensitivity range, and in pure solvents and water. Buffers are often added in published HPLC procedures to sharpen separations, but they only complicate a demonstration and must be washed out before shutting the system down. I provided my own solvents and asked the customer to mix and filter them to my instructions. I used only HPLC reagent–grade solvents and HPLC-grade water. Customers trained in biological techniques love triple-distilled laboratory water. I have often heard laboratory buyers say, "If my water is good enough for my enzymatic reactions, it should be good enough for your chromatography." Wrong! Volatile organics can codistilled with water, and triple-distilled water has killed many otherwise good demonstrations.

The sales representative should research the conditions he or she will be using or have his or her application laboratory find conditions for the demonstration. I almost lost an order because the customer found the conditions in the literature and made up the mobile phases with buffered solvent before I arrived. We could get no separation until I sent him to lunch with the salesman, dumped his solvents, and ran scouting gradients to find usable conditions. His mobile phase contained only 25% solvent; my final separation solvent needed 75% of the stronger solvent. The literature was incorrect, as it often is, because someone had probably mislabeled his solvent pumps and misreported his reaction conditions. The final separation was convincing, as was my ability to solve the problem on the fly, and the order was placed.

6.2 USER TRAINING SCHOOLS

Operator training schools are a great value-added tool to apply to an instrument purchase. The first manufacturer that I worked for as a salesman provided a week-long hands-on training school for one or two users at the vendor's facilities as part of the instrument price; the student's transportation and expenses were paid by the buyer. With another vendor, we offered tuition wavers for two users for a two-day slide presentation school. Buyers could send users to our regional center or, if they had purchased enough instruments at one site, a two-day school would be offered at the buyer's facility. We designed these schools to make the user successful as soon as possible and to ensure user loyalty to the vendor to assist in further purchases.

The value of these schools can be judged by successful publications by our users and by the repeat business they generated for the vendor. We awarded a diploma for completed attendance, and the company received irate telephone calls if the diploma arrived late. The school at the vendor's facility had the advantage of providing hands-on instrument training and removing the students from the distractions of their work site. The school at the buyer's facility was usually combined with an extra day of consulting. I watched and helped instrument users run their own already installed instruments and provided advice on how to optimize their separations. Often I left my instrument operation protocol sheets with them as a reminder of the things I had taught them in the seminar and the consultation.

6.3 VENDOR APPLICATION DEVELOPMENT LABORATORIES

Vendor application development laboratories can be a great benefit to new users. These laboratories publish or make available collections of general methods that have been validated. A trained laboratory specialist can do the methods research if a user needs help in developing a specific analytical method, either as a service associated with an instrument purchase or on a fee basis. Usually if the buyer pays for the service, he or she retains exclusive rights to use and publish the developed method.

Examples of published protocols in the technical literature and in collections from various vendors' methods laboratories can be useful talking points for specific methods for the user's laboratory. They still need to be validated and adjusted to fit the laboratory's needs. The method can be modified enough to separate the user's compounds, even if they are different from the reported target materials, if all compounds fall in the same general category—for instance, separation of phospholipids or purification of proteins.

6.4 TECHNICALLY TRAINED SALES REPRESENTATIVES

Not every instrument company hires a sales force with technical training. Many companies feel that a good salesperson can sell anything from tractors to used cars, and people with training in scientific fields are expensive to hire and employ. The technical sales representative is an additional benefit to the customer that is often overlooked.

My technical training and hands-on experience with the instrumentation allowed me to understand my potential buyers' research problems and work with them by offering guidance on the equipment they would need and approaches to making their separations. Since my training was usually in a different technical discipline, I looked at their problems with different eyes and tended to see the forest rather than the trees. Sometimes I was able to offer simple suggestions that ended up leading to improved publishable separations.

This caused some customers to suggest that my name be added to their publications. I avoided this, saying that my job was to be a confidential consultant or adviser. In many of my commercial accounts, I had to sign a confidentiality document. I told customers that my job was to be the 250-pound bumblebee going from flower to flower, gathering ideas and sharing them when I could do so without violating their confidence. If my name was listed on their papers, it would complicate my job when I went to call on their professional competitors, and publications were not important to the job I was hired to do. Instead, I collected the ideas, built systems where I saw trends, and published them as part of my training schools and the textbooks I eventually developed for these courses.

6.5 VENDOR-SPONSORED TECHNICAL MEETINGS

Many vendors provide financial support and encourage employee assistance for regional and national technical meetings. Some of these meetings turn out to be poorly disguised marketing devices for the vendor's hardware. A few grow to be well-attended forums for technical discussions, platforms for technical publications, and poster sessions for rapid dissemination of technical information.

These meetings are often accompanied by instrument exhibitions that provide multivendor exhibits of the latest and best hardware offerings.

These provide a useful service for potential buyers seeking information on technical specifications; there is nothing like seeing what the hardware actually looks like and talking to people who may have used it, no matter how biased they may be. Sales representatives accompany the exhibits to provide information on the wonderful features and benefits of these state-of-the-art instruments. If you do your homework, you can use probing questions on the last day to sort out the important features at a time when the sales representatives' feet are aching, their backs are sore, and they tend to be a little more honest about their equipment.

6.6 POSTSALES SUPPORT

The support provided by the vendor to ensure your success with the instrument after a field service technician installs it is the true measure of the vendor's commitment to your account. You are dealing with a con instead of a pro if the next time you see a sales representative is when you express an interest in making a new purchase.

Support can be evaluated before the sale only by talking to a colleague who has benefited recently from the vendor's help. Good vendors should offer application support, possibly from an on-line library of methods papers. They should provide information on sample preparation that will ease and speed your analytical methods. Their manuals should provide lists of recommended spare parts and maintenance tips for keeping your system up and running. If things do break or go wrong, the vendor should provide responsive, informed, and cost-effective field service support.

6.7 COST OF CONSUMABLES

Buying an instrument or an analytical system without buying the consumable items that are necessary to operate it is like buying an automobile with no gasoline to run it. Some instruments require few consumable items to achieve their results. A melting point apparatus may need disposable sample tubes, a microscope requires slide plates, and a autosampler may need sample vials with special caps required by the sampling system. Simple vacuum pumps for laboratory evacuations and primary stage vacuums for MS systems must have periodic pump oil inspections and replacements.

The prices of power managements systems such as power strips and surge protectors are often eliminated from purchase calculations. These relatively inexpensive items are critical for long-term successful operation. Electrical Connections break, and the MOS chips that clamp off lightning surges in surge protectors wear out and must be replaced. A vendor of battery-backed power security systems told me that my home city, Saint Louis, averages 168 lightning days a year. Each nearby lightning strike degrades the operating capacity of one of these surge protection strips. I was told by a manager of a large chemical company that in their building expansion programs, "buildings were added like a series of lean-tos, and the electrical system suffered from large surges as people turned heavy equipment on and off." I saw an example of this when the speed sensor on a competitor's HPLC pump burned out, and the pump ran away and literally burned up before it could be turned off or the power cord pulled out of the wall.

The more complex the analytical system, the more expensive are the disposable items necessary for its operation. Chromatography systems require in-line filters, columns, tubing, and solvents for their operation. Chromatography columns can be washed out, but they are usually said to have approximately a 1000-injection lifetime. Mobile phases such as helium and hydrogen purge gases used in GC and SCF systems must be dried; liquid mobile phases used in HPLC must be filtered and degassed before being used. Failure to filter operating solvents will degrade pump check valve performance, plug fittings, and injector loops, not to mention column filters if they get that far. Failure to degas solvents will literally stop some HPLC pumps in their tracks and ruin some HPLC columns by causing them to void. Compression fitting have a finite lifetime before they leak and must be replaced. All of these items have a cost that must be factored into the purchasing decision, usually about 10% of the instrument's cost per year.

Buffers and salts are added to mobile phases to control pH and to improve chromatography by sharpening chromatography peaks. They are a necessary nightmares for system operators because they corrode contact surfaces and plug check valves and tubing lines. Many of these problems can be avoided by periodic pacification with a nitric acid solution following column removal. Check valve repairs, tubing replacement, and pacification are all necessary costs of operating an analytical system. A table of common buffers used in HPLC and LC/MS is included in Appendix B. A protocol for pacification of HPLC systems with nitric acid using a column blank is described in Chapter 9 of my book *HPLC: A Practical User's Guide (2nd ed.)*, listed in Appendix E.

The cost of sample preparation before analysis is often ignored in calculating the cost of the instrument. Samples in solution or extracts from a real-world sample must at least be filtered to remove particulate matter. If the sample is to be injected into an HPLC or GC, the pores of the filter must be smaller in diameter than the HPLC tubing diameters and the

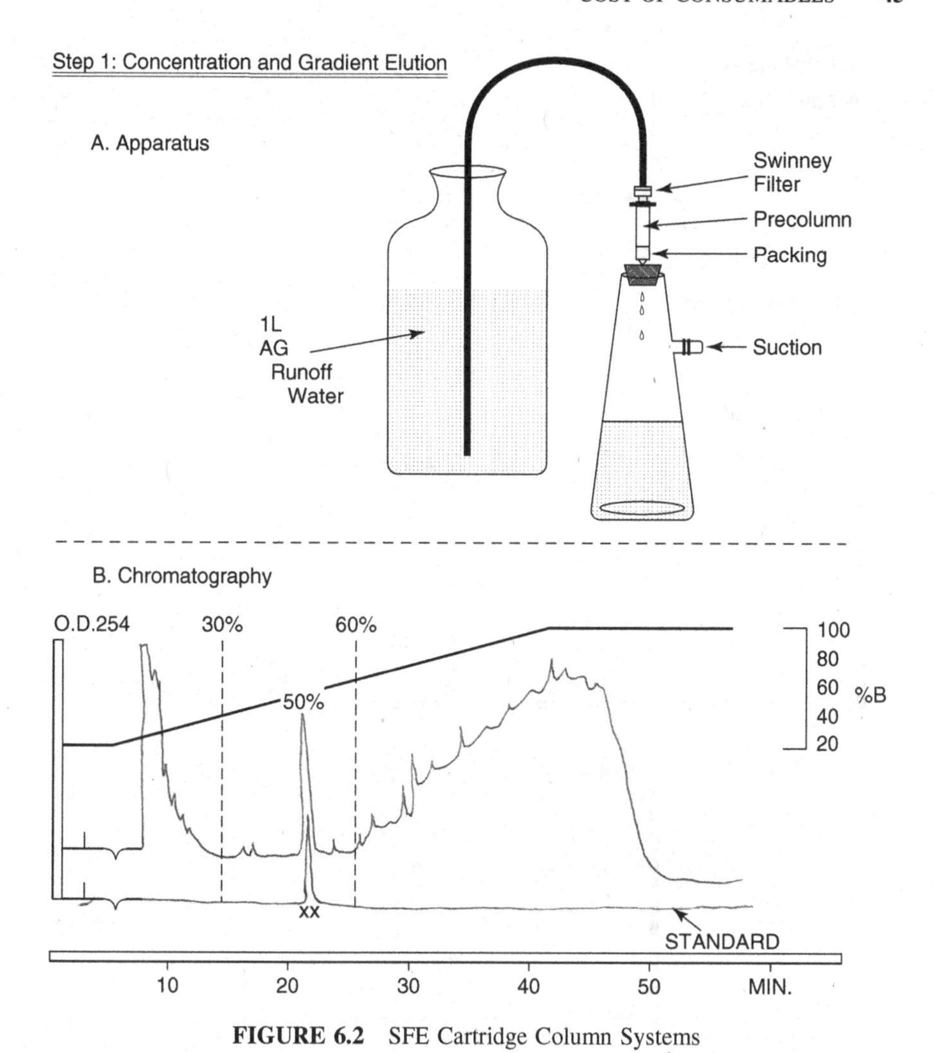

FIGURE 6.2 SFE Cartridge Column Systems

pores of the system filters. Real-world analytical samples are complex, and many impurities interfere with the target compounds for analysis, leading to failure of analysis and decreased sensitivity. Introduction of sample filtration and extraction (SFE) cartridges provides a technique for rapid preinjection removal of polar and nonpolar impurities using a windowing technique described in Chapter 9 of *HPLC: A Practical User's Guide* (see Appendix E) and in Figures 6.2 and 6.3.

Basically, this involves dissolving the sample, diluting it with a weaker solvent, and injecting the solution onto the disposable cartridge column. The cartridge can be washed with the weaker solvent to remove loosely retained impurities, with a stronger solvent mix to remove the sample for

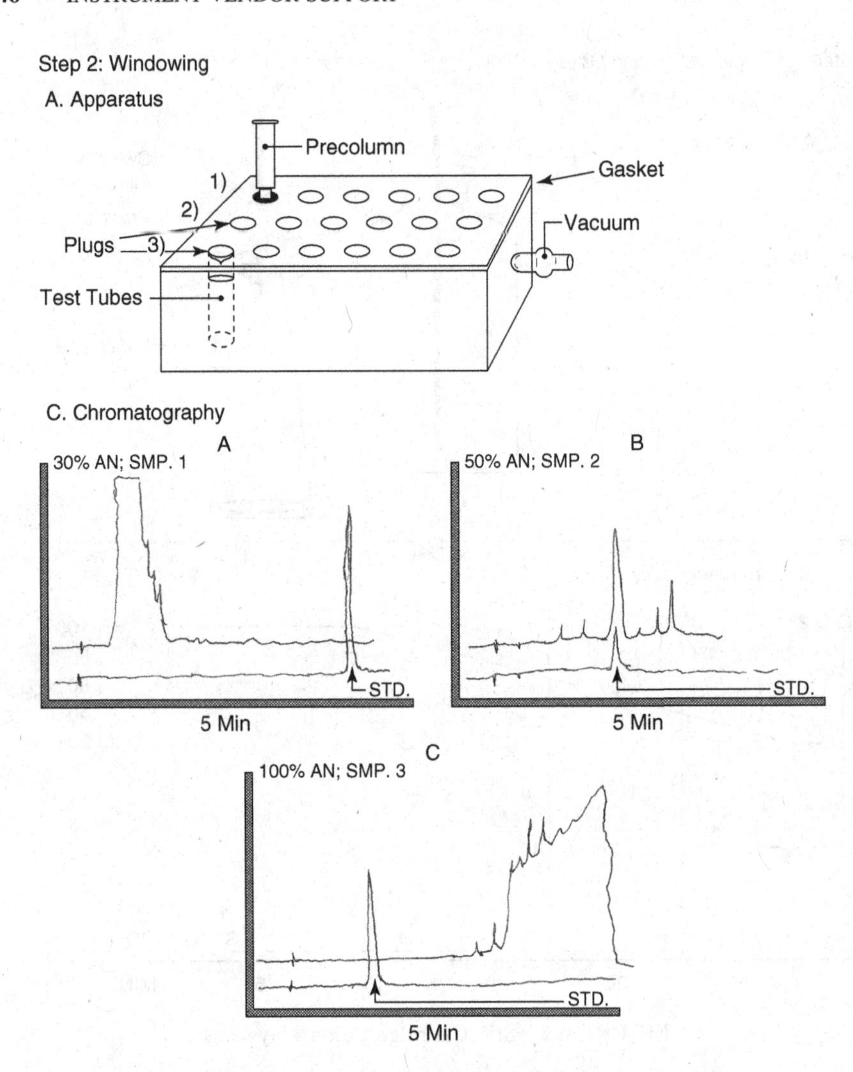

FIGURE 6.3 Windowing with the SFE Cartridge System

analysis, and the more strongly retained impurities can be discarded with the cartridge. The natures of the weaker and stronger solvents depend on the cartridge media. By playing with the composition of the wash for the less retained impurities and for the composition of the sample eluting solvent, the window frames for the conditions to retain and then elute the sample can be adjusted. On a octyldecylsilane (C_{18}) packing, the weaker solvent would be water, the stronger solvent might be acetonitrile, the weak retaining impurities would be more polar than the sample, and the strong retaining impurities would be more nonpolar. A wide variety of packing materials in an inexpensive cartridge column are available, offering a variety of chromatography modes for sample purification.

An SFE cartridge is usually made up of a small amount of chromatography medium trapped between filters in a syringe barrel. A sample in solution can be diluted, placed in the barrel, and either pushed with a syringe or pulled with vacuum onto the packing medium. The sample on the column can be washed with dilute polar solution and then eluted with a stronger solvent. The more nonpolar impurities may be discarded with the cartridge (see Figures 6.2 and 6.3).

Sample purification with an SFE cartridge is not limited to GC, SCF, and HPLC systems analysis; it can be used to clean up samples for other analytical instruments as well. The literature contains reports on the use of these cartridges to purify samples for UV, infrared, nuclear magnetic resonance, and MS. Obviously, if these SFE cartridges will be used in the operating protocol for the analytical instrument, their cost must be factored into the system's purchase price.

7

LABORATORY INSTRUMENT SERVICE

Everything wears out, breaks, and has to be fixed or discarded. Murphy's law accompanies every instrument sale. If you know what you are doing and have the spare parts, you can often make the repairs yourself. Instrument warranties are supplied to protect both buyers and vendors against manufacturing problems early in the instrument's life. Electronic problems usually show up in the first 90 days of operation; hardware problems may not occur for years if the instrument is kept clean and proper preventive maintenance is done regularly. Instrument problems serve a secondary purpose of making the buyer aware of and dependent on the manufacturer's service department.

7.1 QUALITY IS JOB 1, QUALITY SERVICE IS JOB 2

The key to excellent instrument operation is a well-constructed instrument. The instrument will continue to perform flawlessly if it is serviced regularly when the design is correct and manufacturing is done with precision parts and attention to detail, Time is money even in a laboratory heavily stocked with free labor in the form of postdoctoral and graduate students. The more the instrument is available, the more it can be used to produce useful experiments and generate research papers, the life

Buying and Selling Laboratory Instruments: A Practical Consulting Guide.
By Marvin C. McMaster

blood of the university professor, or results and invoices for the cost-per-test laboratory. Available and prompt service is a critical component of the vendor's reputation when a instrument purchase is being considered. Turnaround time for a spare or replacement part is also critical. Overnight package delivery is useless if the manufacturer's inventory is too low or the company cannot respond promptly to purchase orders.

I experienced a serious service problem with a major account. This customer's company purchased 80% of the instruments sold in my territory. The manufacturer I worked for had maintained a service man in the same city as their research laboratory who was respected by the customer for his ability and his prompt service. He was so well respected that the manufacturer promoted him out of the territory and then failed to replace him for nine months. In the meantime, his customers were to be serviced by other service representatives flying in from other cities.

Since we were in a very competitive sales situation, I realized that this changing service situation would endanger my ability to maintain sales control of the account. I pulled the literature out of my catalog case, stocked the bag with spare parts from my demonstration inventory, and carried it in the trunk of my car. If a customer had a minor problem such as a broken HPLC pump plunger, a leaking seal, a bad analysis column, or a clogged flow cell, I fixed it by using the parts in my catalog case. I then told the customer to order a new part and replace mine when the shipment came in.

The customer thanked me for getting his system up and running by referring me to three colleagues who were getting ready to order new systems. I kept my management informed on what I was doing. They were unhappy that we were losing the billing time for the service calls but were thrilled that I was exceeding my sales quota. I gained some very interesting information from this process. Most service problems are minor but very detrimental to the customer in the form of research downtime. If the service problem was beyond my ability to solve, the customer often realized that it was serious, was usually willing to wait for the service professional to fly in, and tolerated the higher billing more easily. I tracked these service calls and did everything I could to ensure that the problems were handled promptly.

My time as a sales representative and serviceman had other benefits. I was able to develop a series of rapid column and system cleanup techniques that increased customers' running time and made their systems and columns last considerably longer.

7.2 SEPARATING INSTRUMENT AND APPLICATION PROBLEMS

Taking a few minutes to think about what is going on and what might be the cause of an instrument problem before you call the service technician

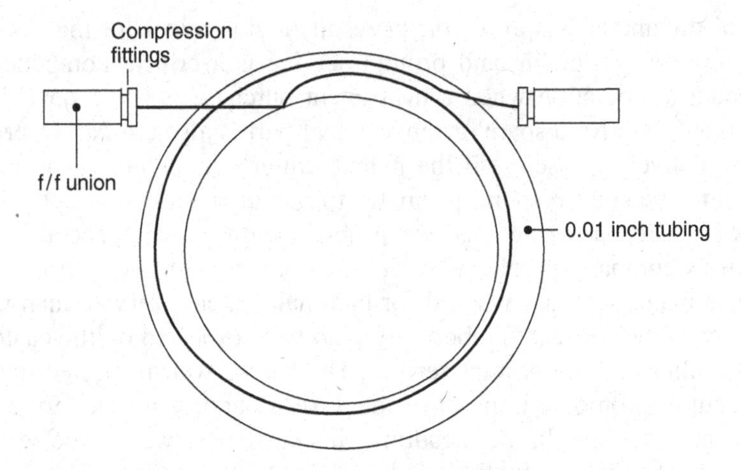

FIGURE 7.1 A Column Blank

may save you downtime and service billing costs. I found that service people were only trained to repair electronic and mechanical problems. If the problem involved the method or the column chemistry, they were often unable to recognize it and spent much work time and billing time trying to solve it. I developed the concept of a *column blank* to simplify the problem to one of electronics and hardware so that the service person could treat it (see Figure 7.1).

The column blank is made from 5 feet of 0.009 mm tubing with attached tubing and fittings bypasses. It replaces the analytical column while providing back pressure allowing the pump's check valves to open and close.

The majority of problems in HPLC systems are due to column problems, and most of these are due to bad water or changes in the column surface characteristic due to accumulation of impurities. My first attempt to solve this problem involved removing the column and replacing it with a virgin column that had only been used with standards and thus eliminated the problem of column chemistry changes. The column blank, which bypassed the system column, completely removed column chemistry from consideration, provided enough back pressure to allow the system to operate, and was much less expensive than supplying a virgin column and standards for each service person to carry for diagnosis. Additional advantages of the column blank emerged with use; the blank was impervious to strong oxidizing acids and could be used in system washout for cleaning pump check valves and detector flow cells, extending the system's operating lifetime and reducing wear on check valves and other components. The injection response with a column blank was almost instantaneous compared to injections through a column, again proving an aid to speed the diagnosis of hardware and electronic problems.

7.3 REVERSE-ORDER DIAGNOSIS

Another tip learned while I was wearing both a service and a sales hat is that there is an optimal order for solving service problems. The method normally used I call "throwing mud on the wall." Diagnosticians identify a module in the system as the culprit, usually by consulting their magic ball, declares it to be the source of all of their problems, and want it replaced or fixed. This wastes time if the module is not at fault.

It is better to approach this problem systematically. Start at one end of the system. Prove that the first module is working well and then use that module to prove that the next module is working. I usually started at the back end of the system and proved that the strip chart recorder or the data system was running and deflecting in the expected manner when isolated from the rest of the system. Next, I connected the detector to the strip chart and showed that it was producing a signal on the strip chart. I looked for electronic and power line glitches while running the strip chart's bed. If there seemed to be none, I increased the detector's sensitivity and looked for a change. Once the detector was tested, I moved to the next module, with a column blank in the system replacing the column. I shot standard through the injector and looked for peaks in the baseline. The last candidate for investigation was the pumping system (see Figure 7.2).

The same technique will work with any modular system and also with any instrument that can be separated into isolated components. Start with the data output module and work toward the sample input. Try to diagnose electronic problems before you move to sample handling. I once had

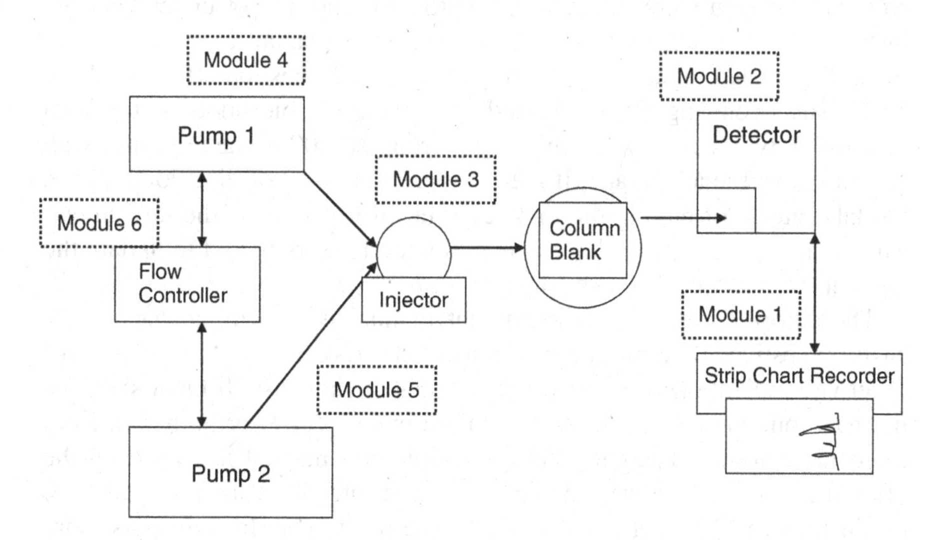

FIGURE 7.2 Reverse-Order System Diagnosis

a detector problem that seemed to be a case of electronic degradation but turned out to be only one of sample decomposition on a flow cell window under UV radiation. We finally traced the problem to a slow-eluting compound leaching off of a chromatography column; washing the column and the detector flow cell with a strong solvent solved this problem.

The point is that service done regularly by an informed service professional increases the value of your instrument purchase by providing consistent operation that will impact your output of results. Your ability to understand your research systems and protocols will optimize the results you obtain from that service.

7.4 SERVICE RESOURCES

The best service is instantaneous service. When an instrument stops working, the best response is to pull out the manual, look for diagnostics, simplify the system, run through the troubleshooting tips, and get the system back into operation.

Instrument manuals are rapidly disappearing from the instrument environment. Often the manufacturer will ship a very short Read-Me-First document and then put all the instrument description, troubleshooting, and diagnostic aids as a lookup on their web site or in the Help section of the instrument's computer software. Most Help sections are written by software programmers who cannot write and who do not understand the instrument's hardware or its problems. A good manual containing hardware diagrams, diagnostic tips, frequently asked questions (FAQs) about previously occurring problems, and systematic troubleshooting steps can be absolutely essential when things go wrong at 1:00 in the morning. Web sites do have the advantage, if there happens to be an on-line computer in the laboratory, of having the most current information on the instrument. This assumes, of course, that the manufacturer has bothered to update the contents of the web Help screens.

The completeness of the instrument manual is something that can be easily investigated before the instrument sale. Ask the sales representative to let you look at one before you bid or make a purchase. If he or she does not have one, it can be obtained from the manufacturer or borrowed from a recent customer. The care that a manufacturer has put into creating the printed manual or the web manual often indicates the care and quality of the instrument. It is very instructive to see if the manufacturer puts more effort and quality into the marketing literature than into the manual.

The same type of information can be obtained from an investigation of a manufacturer's training school, if one exists. Get information on the curriculum or a school schedule outline. Obtain the name and phone number of someone who has recently attended the school from your sales representative, call this person, and get his or her impressions. Also inquire about this person's startup experience, the help received from the company during startup, and the service experience. If this person has had no experience with the service department, that is not necessarily bad; it may mean that the instrument is a quality system that keeps working. This may seem like a lot of trouble, but it will pay off in instrument performance quality on your research problems.

7.5 SPARE PARTS INVENTORY

Find out where the wear points are on your instrument, the things that need to be replaced on a routine basis, and the preventive maintenance that can cut downtime. Get information from the sales representative, from his or her service representative, and from experienced customers on the sales person's referral list. This information will lay the groundwork for a list of needed spare parts that should be kept available for routine service. Keep a list of needed spare parts, update it as items are drawn out and used up, and reorder as soon as possible. Nothing slows laboratory production more than having a wear item break down and lacking a replacement because someone failed to order it.

Wear items on an HPLC system are filters, pump plungers and check valves, analytical columns, plugged tubing and leaky ferrules, contaminated analytical columns, detector lamps and flow cells, and recorder paper, pens, or printer cartridges. None of these take more than 15 minutes to replace if you have the spare part and the correct tools.

Lock up your tools! Unlocked tools have a very high disappearance rate. You can keep equipment up and running more consistently by making someone in the laboratory responsible for checking in and out tools and spare parts. It is better to use a senior technician than a graduate or post-doctoral student; technicians are usually more responsible and have a lower turnover rate.

7.6 DIAGNOSING GROUNDING AND STATIC PROBLEMS

The most frustrating and expensive service problems are strange, transient glitches. When an instrument fails completely and stops running, it can usually be repaired or replaced fairly easily. It is the occasional problem that is most frustrating.

Usually the first step in problem solving is to simplify the system as much as possible. Disconnect accessories and get down to basics; once you have checked the basic system, you can start adding back accessories one at a time until the problem recurs. If possible, make sure that all the components are plugged into a common power strip or surge protector. Ground loops from components on separate power sources can lead to glitches ranging from static problems to baseline drifting, spikes, or shocks.

Check the manual for diagnostic tips that allow you to check a system's performance. On chromatography systems, this might be a stable injection standard solution that can be injected through a column or a column blank to check system performance. Run your diagnostics when a system is new. Record the results in the manual and check against this performance when problems occur. This technique not only works for chromatography columns but can also be used to check detector lamp performance. Talk to the service technician and find out the diagnostic setting he or she uses to check equipment. Technicians usually have a set of extreme settings and a minimal standard that must be met before they recommend replacing a lamp or a detector. Find out what they are. I once asked a service representative that question about an MS detector. He said that the technicians were told to maximize the repeller setting, run the EM voltage setting to 3000 v, and see if the 502 mass peak of calibration gas exceeded 10% of its initial value with the 69 and 131 mass peaks calibrated.

Some of the most annoying problems to diagnose are static discharges that often occur during the winter months in very dry atmospheric conditions. They can stand your hair on end and produce nasty shocks. When the same thing happens to a poorly grounded instrument, it can reset instrument values; I have heard of one case in which it wiped the system's microprocessor, effectively lobotomizing the instrument. Some of these problems can be avoided by plugging all components into the same power strip. I have had customers who had to wear a grounding strap on their wrist in order to be able to touch a recording integrator that we sold for their chromatography system.

It is important to know that the MOS chips that clamp off lightning surges in surge protectors wear out in time and the surge protector needs to be replaced. Most of these strips have a light indicating that they are still working. If you have a large number of yearly lightning strikes in your locale, you might want to accelerate your replacement time. At least one of the static problems I have seen was attributed by an MS user to a lightning strike that burned out the surge protector on his system. We were able to stabilize the detector signal on the system so that it could be calibrated by running a grounding strip from the detector to the chromatography portion of his LC/MS after plugging all components into a common power source.

8

RECYCLING THE SYSTEM

What do you do with a research system that has reached the end of its usable life? It can be upgraded with newer, more powerful detectors and data processing systems. It can be automated into a dedicated hands-off analyzer for routine nondemanding analysis. It can be set aside as a practice instrument for training new researchers or technicians. It can be donated to a worthy university researcher as a research and teaching instrument for a tax donation. It can be sold to an instrument reseller. Finally, it can be used as a boat anchor or for metals recovery.

8.1 THE DEDICATED RECYCLED SYSTEM

When I was selling HPLC equipment, I ran into an interesting illustration of how to extend the working life of an HPLC column that can be used to give directions for extending instrument lifetimes. Most clinical labs use an estimate of 1000 injections as a typical lifetime of a C_{18} analytical column. For a single-shift-per-day operation, a column might last for about a year before a replacement column is needed. The customer I was talking to at a major midwestern medical research clinic said he typically got a four-year lifetime for his C_{18} columns.

Buying and Selling Laboratory Instruments: A Practical Consulting Guide.
By Marvin C. McMaster

I had worked out a number of column washing and diagnostic techniques for extending column lifetimes, but these required downtime use of an HPLC pumping system and a detector that most laboratories running cost-per-test analysis were reluctant to give up. By the way, a column cleanup HPLC might provide an occupation for an obsolete dedicated system. But I would never expect column cleanup techniques to quadruple a column's lifetime.

When I asked the customer to share his method for extending his column lifetime, he said it was very simple. He found that he ran four dedicated methods, each more demanding that the one before it. He would apply a new column to the most demanding separation. When injection of standards showed that the separation was failing, he would wash the column out with six-column volumes (about 20 mL) of a strong solvent and move it to the next most demanding separation. When this separation also failed, he would wash out the column again and move to the next separation. After the column failed on all four separations, he would have a technician give it a full wash with all of my washout techniques to see if any activity could be restored. If not, the column would be partially emptied from the exit end to provide clean packing material for guard columns or for repairing the tops of columns whose bed had been dissolved or degraded. Using these techniques, he could get four analytical application passes and some good repacking material from every column, the equivalent of using all of the pig except the squeal.

Every instrument system, like an HPLC column, has a lifetime. Instrument technology advances leave older systems behind in the movement to faster, easier-to-use systems and systems with more specificity, higher sensitivity detection, and more detailed reporting. Sometimes existing systems can be upgraded with newer software, faster computers, and more advanced detectors to stay current with the state of the art. But eventually, a point will be reached when nothing you do will be sufficient to make the older system meet more stringent research requirements. At this point, the temptation is to give up and cast the system off in some form to recycling.

Before we write off the system's cost and throw it out of the laboratory to recover the bench space and the operator's time, let us see if there is some way to extend our investment's life. For instance, most of a multipump HPLC gradient system's cost is in the price of the pumps. An isocratic dedicated HPLC system requires only a single pump. If we add another injector, a relatively inexpensive detector, and another A/D data card to the data acquisition computer, we can create a pair of parallel isocratic HPLC systems to run two dedicated analyses. The gradient controller can be retained as a pump controller or a timer for diversion valve control; if need be, it can be used to re-create the original gradient system with two in-line detectors for methods scouting, as illustrated in my book *HPLC: A Practical User's Guide (2nd ed.)*.

There are often many jobs that a system can do in the laboratory if it is used as a dedicated analyzer. Look at the jobs and projects that are being considered for implementation by various research groups that will not need the full power of a state-of-the-art instrument.

I know of a former HPLC dual-pump gradient system that was separated and converted into a protein separation system and a protein analysis system (see Figure 8.1). One pumping system was used to run a protein size separation column and the HPLC detector into a fraction collector. The other pump and a second protein size separation column was made into a protein detection system with a couple of obsolete fixed-wavelength detectors with their wavelengths set at 280 nm (histidine groups) and 254 nm (phenylalanine/tyrosine groups) and a dual channel strip chart recorder. The gradient system was 12 years old when it was converted and was considered by the laboratory director to be "a real boat anchor." It was rescued and pressed into service for protein separation and analysis by graduate students and was still in full-time operation when I was last in that laboratory 20 years after the original system was purchased.

8.2 TECHNICIAN TRAINING INSTRUMENTS

Everyone starts their laboratory career with the same lack of hand-on experience. Instrumental analysis courses tend to be very theoretical, and generally their laboratories use simple, obsolete equipment. Analytical instruments are expensive and usually are dedicated to the laboratory's major work. Freeing state-of-the-art systems and trusting them to inexperienced operators is just not done. Obsolete instruments can be freed to serve this training function and provide almost the same experience that newer ones provide. However, giving a new technician an old instrument, a manual, and some samples to analyze is seldom very effective.

I ran into a similar problem when I was selling instruments. A customer would place an order for a new system, it would be shipped to the laboratory, and if I was lucky in my company selection, a service technician would come in, connect the pieces, make sure that the electronics fired up, and left the account with an instrument that no one knew how to use. An unused instrument taking up lab space quickly became a negative force preventing new orders from the account. To avoid this situation, I arranged to get into the laboratory as soon as the instrument was installed to make sure that the customer turned it on and began using it.

This turned out to be a problem. The principal investigator, his senior technician, or a postdoctoral student had purchased the instrument. They

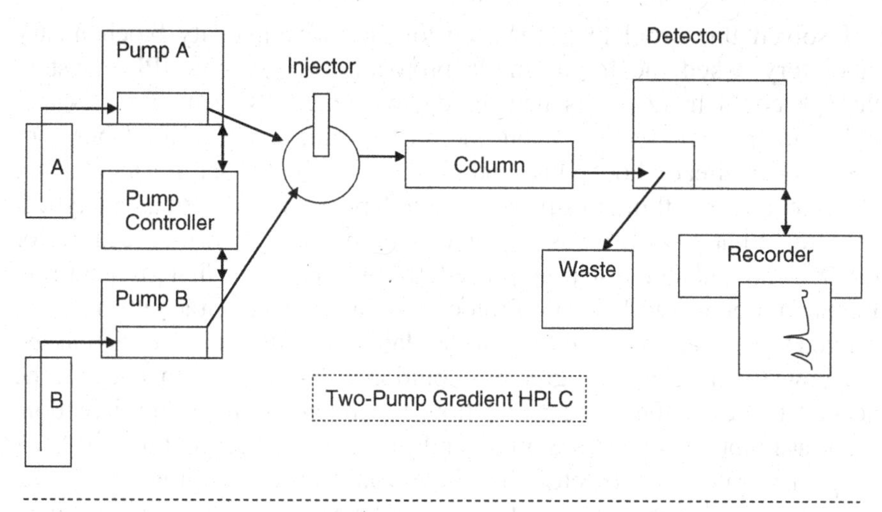

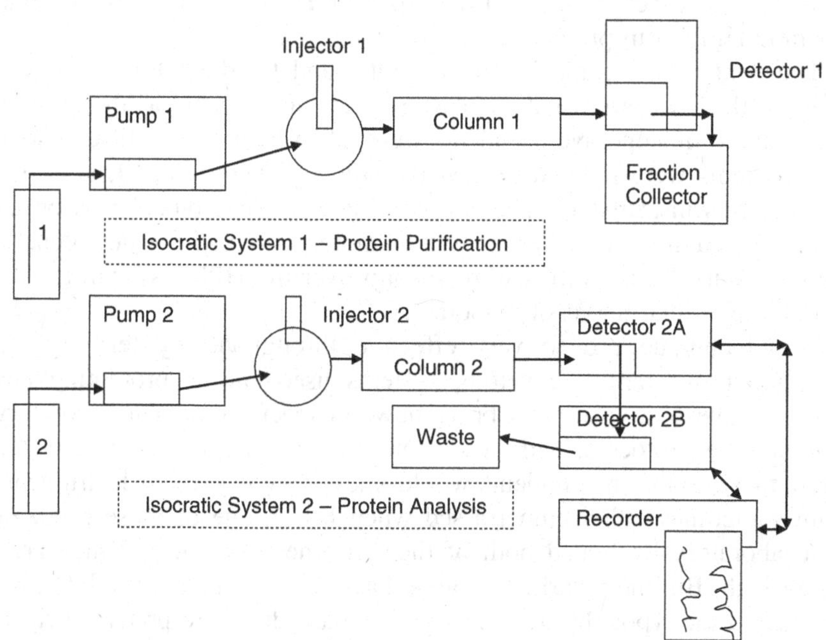

FIGURE 8.1 Recycling: Gradient HPLC to Two-Isocratic Systems

were the ones who turned up at the laboratory to learn how to use it or were assigned to go to the training school. The day-to-day operator seldom had enough status to be selected for the training. Even worse was the case of an instrument purchased for a newly hired technician; unless this person came from another laboratory with the same instrument, he or she seldom had the required training and was forced to use the new system and its manual to teach himself to use the instrument.

I solved this problem when a customer in a university biochemistry laboratory asked me for a simple protocol for his new HPLC system that he could hand to his new graduate students. I went home, sat at my computer, and thought about the problem for a while. The laboratory system was already hooked up and checked out by the service technician; at worst, thc only thing this pcrson would havc to do was add an analytical column. What I needed was a simple isocratic step-by step procedure for the first run and then a step-up procedure for making the first gradient run. I created a one-page Isocratic Protocol and a one-page Gradient Protocol, printed them out, and gave them to the laboratory director. He then asked me how to make up the gradient solutions, and wanted to know if he needed to degas the solvent and how to prepare samples for injection. Back at home, I created a Sample and Solvent Protocol and took it back to the laboratory. I expected that they would place the protocols in the system manual (see Figure B-4 in Appendix B for a copy of the original isocratic HPLC run protocol).

The next time I came in, the Isocratic and Gradient Protocols were taped to the wall, protected by plastic sheets, and treated as though they were carved in stone. New students were taken before the HPLC system, the protocols were pointed to, and the students were told, "This is how you run this instrument." The Solvent and Sample Protocol was on the wall, covered in plastic, over the sink in another room where they prepared their solvents. Those sheets were still up over the HPLC system the last time I visited that university laboratory.

This technique proved very effective throughout my territory and throughout my career in getting systems used and in promoting new sales. I have had customers bring new prospects into their laboratory, point to the protocols, and say, "You want to buy from this guy. He wrote these so that my students would know how to run our instrument." I am sometimes rather embarrassed when I seek one of those protocols in a laboratory. I created both of them in one evening, and they are a little rough. But they work! I've gone back, cleaned them up a little, and eliminated the typos, but the ones on my hard drive are pretty much the ones I handed out that day. The basic HPLC protocols were upgraded and included as figure B.4 in the Appendix and in a more complete form in *HPLC: A Practical User's Guide (2nd ed.)*.

8.3 UNIVERSITY INSTRUMENT DONATION

If there is no place in the laboratory for additional dedicated analyzers or training instruments, consider offering the instrument to a local university or technical training school as a tax-deducible donation. University chemistry and biochemistry departments are always short of funds for purchasing teaching instruments and could use an extra analytical instrument for research or for training students. You will need to check local regulations for tax-deductible donations to such institutions to see if the goodwill generated would justify the time and effort involved in setting up a donation. A donation usually creates a dynamic flow of creative ideas between institutions, and the University Laboratory can act as a local source of technicians and new scientific staff at a later time.

8.4 USED-INSTRUMENT RESALE

Consider offering obsolete equipment for resale if instrument donation appears to involve too much effort and distraction for too little return. Technical magazines and *Chemical and Engineering News* often contain ads from instrument resellers or of instruments for sale. Consider advertising, but also consider who bears the cost of shipping and reinstallation.

There are three things to consider in reselling instruments:

1. Be sure that the data on the computers will not be needed at a later time. Cost-per-test analytical laboratories may need these data as a defense in later lawsuits.
2. Hard drives need to be reformatted before disposal, with data overwritten with zeros to protect them.
3. Do not expect a large financial return on old equipment. The reseller usually realizes the only real profit made, but reselling is environmentally responsible, benefits the final owner, and is cheaper than having the system hauled off by the garbage collector.

8.5 METAL RECYCLING

The final fate of many obsolete systems is to be stored and forgotten in some out-of-the-way location in the institution. Many university and corporate facilities are filled with old instruments stored in basements and hallways. The rationale is that the equipment in these elephant burial grounds might be needed at some future date. This is very irresponsible and simply shifts the problem to someone else at a later time.

Check to see if your company has worked out a relationship with a company or a school that does metal recovery from electronics and computers. Again, check the magazines that contain information on instrument resellers; they may provide references to metal recovery companies.

Also, investigate computer technical magazines; the disposal of television sets, monitors, system modules, cables, and computer cards is a growing problem and is slowly being addressed by strippers and recycling organizations. The conversion to high-definition television broadcasting is creating a landslide of old analog televisions that need to be recycled, and is accelerating the availability of recyclers and lowering the cost of doing this recovery. Besides the iron, steel, and aluminum in instrument cases, there are a variety of more expensive metals such as titanium, gold, and silver that help defray the cost of disposal.

If all else fails, I know of at least one HPLC manufacturer that used an old HPLC pump as a boat anchor. This was not a very environmentally responsible solution, but it seemed to serve its purpose fairly well.

PART TWO

A GUIDE TO THE SELLING PROCESS

9

BUYING RELATIONSHIPS

The key to making a successful purchase is to find a professional salesperson or an in-house instrument guru who can assist in selecting exactly the equipment needed for the customer's work. Such a salesperson can comprehend the technical aspects of the instrument needs and can help define and write bidding specifications that let the buyer obtain the correct system. Such a buyer can negotiate a fair price for the instruments and supplies that will be needed to operate the system.

Every instrument purchase should be a win/win relationship. Every salesperson should be in your laboratory to help you solve your analytical problems. A professional salesperson will ensure that you receive exactly the equipment you need to complete your analytical tasks at the lowest price so that his or her company makes a profit sufficient to stay in business. In other words, both you and the company should equally meet the customer's needs.

In the real world, few win/win purchases are made. Most instruments and systems are purchased at list price and are more elaborate than the needs of any possible research project. Or worse, they are bargain basement systems that barely meet the needs of the customer's current projects and lack expandability to solve new problems. In the worst case, the instrument simply fails to serve the purpose for which it was purchased and sits on the shelf or laboratory bench as a glaring warning for the next purchase.

Buying and Selling Laboratory Instruments: A Practical Consulting Guide.
By Marvin C. McMaster

9.1 WIN/LOSE SELLING RELATIONSHIPS

We have all been involved in a purchase that we have come to regret. Think about the shoes you bought that were too tight that a salesperson assured you would stretch to fit your feet or the suit that looked good under shop lights but bizarre in daylight. You can remember the exact price of the suit that was slightly too tight, but you probably have no idea of the cost of the car that ran like a top until the wheels practically fell off.

In an I win/you lose purchase, only the seller benefits from the sale. The salesperson decides on the system and options to be sold, sets the price, and tries to manipulate the sales process to ensure that the buyer purchases the product offered at the listed price. No attempt is made to find out what the buyer needs or to allow any meaningful price negotiation. The techniques used by used-car salespeople are typical of this win/lose technique. The salesperson has a target product to move, usually the deal of the day, and the sale is based on a high-speed description of a string of *desirable features* possessed by the target item. The salesperson stresses the urgency of the discount, which exists only until the end of the day or week or month. Any further purchase discount has to be approved by the manager, and usually involves removing options at a lower price or adding low-priced options at the same price. Pricing games often involve bait-and-switch changes to a less desirable, less capable model.

What has any of this to do with buying laboratory systems? Salespeople with exactly the same type of sales training, experience, and motivational techniques are often used to sell laboratory equipment. The sales techniques may not be quite as blatant, but the intent is the same.

I knew a salesman with a major instrument company who used to sell by walking into the customer's laboratory lugging his company's latest instrument on his hip, dropping it on the laboratory bench, and saying, "Here, try this out. You are the first one in your company to get to look at this. It has all the latest bells and whistles. If you like it, I'll let you buy it as a demo system if you get me a purchase order by the end of the week." If the customer did not buy, he simply picked up his product, took it to the next customer, and said exactly the same thing. He usually inflated the manufacturer's price by 10% and then gave the customer a "deal" of a 10% demonstration discount. This man made a sales career out of carrying around instruments, no matter how heavy, and selling them using this technique.

Once you bought the system it was yours. If you had a problem, the salesperson did not know how to run the instrument or fix it, and he was too busy selling other dropoff instruments to follow up. You might be able to get a service technician to visit by calling the 800 number in the instrument manual. The next time you saw the salesman was when he walked in with another instrument on his hip.

This is called a *puppy dog* sale. Owners of pet stores used to tell people, "Not every family can adapt to having a pet in the house. Take it home and seek if it fits in your household." If you let your family get used to the new puppy, it was impossible to take it away from the children and return it to the store.

Most win/lose salespeople are a little more sophisticated than the one I have described. They have been drilled on the features and benefits of the various systems and equipment sold by their company. They usually start with the top-of-the-line systems and equipment they offer at full list price and sell their way down to less expensive systems if the customer objects to the price. They look for the customer's reaction to one or more of the features offered and then close on the customer-approved benefit, emphasizing how quickly they can get the unit into the laboratory and up and running. If they encounter resistance, they offer to find out if any demonstration instruments are available at a discount; and surprisingly, there always seem to be a few of them.

Laboratories that buy only hardware systems at the lowest price may encounter problems: the instrument will not perform the analysis required or it is inflexible when they need to change the analysis protocol. They may face unexpected costs when they need help to change the method or when the manufacturer cannot supply timely service. Hardware-only buying decisions are the single greatest source of unusable laboratory *white elephants*. I have actually seen a university laboratory buy an obsolete demonstration HPLC system without a detector that was incapable of running the laboratory's desired separation. They were sold this system with the assurance that the laboratory could put the detector out on bid during the next bidding season to create a usable analysis instrument.

The best of these win/lose salespeople may inquire if you know anyone else who is looking for a similar system and ask for a referral. They usually will offer to get you bidding specifications from their company if you need to put the system out on bid. Once you buy, you may see these salespeople again if their company tells them that you have responded to a reply card in a technical magazine or on a routine follow-up visit to your company to ferret out new sales possibilities.

9.2 WIN/WIN SELLING RELATIONSHIPS

Professional instrument salespeople are rare and valuable individuals. They have been trained to find out what the customer is planning to do with the instrument, match the features of their systems to the customer's needs, allow for changing needs by the customer's laboratory, recommend the system(s) that provide the best fit, and help the customer buy. The best of these salespeople considers themselves to be the customer's consultant, helping to solve the laboratory's problems. They understand that the word *sales* comes from the Norwegian word *selye*, meaning "to serve." They usually have enough technical background to understand the customer's application and a willingness to ask questions and let the customer educate them about the problem to be solved so that they can help provide a complete solution. Once the customer agrees that the system will fit his or her need, the salesperson will supply specifications and often will help with the bidding process to finalize the order.

A win/win salesperson understands that the instrument sale is only the first step. The system must be installed, and the customer must begin to use it in his or her research. The salesperson should know how to use the instrument on simple problems that the laboratory personnel can repeat to gain familiarity. Nothing is more detrimental to a future sale than an installed instrument that everyone is reluctant to use.

Applications support from the instrument company may be needed to provide a method protocol so that the instrument can be used to solve the customer's problem. Instrument service may be required if problems occur during startup and in continuing operation. The salesperson inquires about other investigators in the company who might need similar problems solved and stays involved with the laboratory's progress to intercept problems and uncover new equipment needs. His or her role is to become a valued account resource for solving problems by providing state-of the-art systems, application support, and advice.

9.3 BUYING HARDWARE, SERVICE, AND SUPPORT

With instruments more complex than a water bath or a scale, there are three considerations that affect the successful operation of the system: hardware and supplies, service, and support. Most buyers realize that they need to buy hardware that fits their needs and their budget. Most of them know that complex computer-controlled systems require operating and data processing software. They also know that consumables are needed to carry out their analysis, such as solvents, volatile gases, separation columns, reagents, and sample containers. Generally, they have already budgeted for these items.

Less obvious are the service and support options. If the instrument breaks during operation, somebody will have to repair it. If the manufacturer does not offer readily available service personnel, the customer will have to fix it. Large corporations and universities sometimes train and maintain their own service people, but no one knows the instrument as well as the people who make it. When deciding whether to purchase an instrument, it makes sense to buy from a company that can provide in-house service or at least train your service people.

The same applies to application support. If you buy an instrument from a company that cannot show you how to use it to solve your problem, you are the application support. Suddenly, your new instrument has become a research project instead of a research solution. If you buy from a company that offers a training school for operator and application development laboratories, you have purchased a research solution.

But service and support come at a cost. There is no such thing as a free lunch. If you want the new instrument to be truly useful, you must build the cost of service, training, and support into the buying decision. Not all companies offer all of these components, so this information should be acquired in gathering specifications for the purchase decision and built into the bidding specifications.

Price becomes a factor in this relationship only if customers do not feel that they are getting value for their money. A fair vendor profit is important so that the company can do the research and development to provide state-of-the-art updates and new systems, responsive service, and technical support as investigational directions change. And someone has to pay to maintain the sales consultant and make him or her available when this person is needed again.

Laboratories with trained personnel familiar with similar types of instrumentation can often afford to worry only about hardware and service. This is especially true of cost-per-test laboratories that analyze a fixed set of protocols on a regular basis and only rarely add a new analysis to their package of analytical offerings. I worked with environmental analysis laboratories that ran contract laboratory program (CLP)-type analyses where the protocol was specified by the government. The Environmental Protection Agency (EPA) had developed the methods that they were required to use. The laboratory's major considerations were instruments that worked and the maximum number of tests they could carry out in a given period of time.

9.4 ADVANTAGES OF A PROFITABLE VENDOR

A fair price for an instrument or a system includes the cost of manufacturing the equipment, the cost of selling the equipment, and a profit for the manufacturer large enough to allow them to provide service, support, and state-of-the-art instruments in the future. The cost of manufacturing is typically about 40% of the selling price. The balance of the selling price is for marketing, advertising, salaries, benefits, shipping, and a profit for the manufacturer, some of which is reinvested in research, development, and customer support.

The customer wants the manufacturer to be profitable. A manufacturer who is not profitable will go out of business and will not be there the next time the customer needs service, support, or a new instrument. The world is full of orphaned instruments whose manufacturer is no longer in business when the system breaks down and needs replacement parts or service.

9.5 GETTING WHAT YOU PAY FOR AND NEED

In an ideal win/win relationship, everyone will get what they want from the sale. The manufacturer will obtain a profitable sale and an ongoing relationship with a customer who potentially will buy again. The customer will obtain an instruments to help solve his or her research problems, leading to publications and training for the laboratory personnel. A clinical laboratory will obtain an instrument to carry out analyses to solve problems and generate revenue.

People do not buy drills because they want drills; they buy drills because they want holes. If a laboratory instrument fails to work correctly, no one achieves his or her goals and customers do not get what they intended to purchase. When this happens, the relationship is broken dramatically, affecting future instrument purchase. Both the customer and the salesperson's need to focus on long-range effectiveness.

Single-instrument sales without a prospect of future purchases are expensive and stressful. Each sales experience is time-consuming and expensive. Both the customer and the manufacturer need to ensure an effective and productive buying relationship that leads to customer satisfaction. This is usually achieved by follow-up by the field sales and service people to ensure that communication with the customer is maintained and that problems are detected and solved as quickly as possible. Nothing sours a relationship more than a problem that is allowed to fester. Nothing builds a relationship and creates future sales opportunity better than attention to what is occurring in the customer's laboratory.

Look for a professional salesperson the next time you buy an instrument—one who is interested in what you plan to do with it, who asks questions and listens to your answers, and who makes sure that what he or she is selling you will fit your work. Find out how he or she has helped other people in your institution or on his or her referral list. Find out if this person makes sure that the customers were successful in using their new purchases. Verify that the sales were win/win events both for the salesperson's company and for the customers.

10

SALES JUSTIFICATION

The reasons for buying a particular item are often misunderstood. Most buyers feel that they consider all their needs for the purchase, and the amount of money available, and make the most *logical decision* based on these facts. Salespeople are told that customers do not make logical decisions to buy; they make *emotional decisions* and support these decisions with logical justifications. When I first heard this statement, I became very angry because I consider myself to be a logical person. Almost everyone to whom I explained emotional buying had the same reaction. However, it's hard to argue with success. I have made very logical sales presentations, and no one bought. I have made bright, emotional, pictorial presentations and made sales. The salesman was the same; only the approach changed.

10.1 EMOTIONAL DECISION MAKING

The eyes provide logical access and the ears emotional access to the brain. Salespeople are taught to sell to the ears and to provide pictorial (emotional) reinforcements through the eyes. If cars were purchased logically, they would all be black and no one would be attracted to that new-car odor. We buy because we like the way they look and smell. We like the

Buying and Selling Laboratory Instruments: A Practical Consulting Guide.
By Marvin C. McMaster

feeling of power when we drive them. We buy and then acquire a long list of features to justify our purchase decisions to our spouses and bosses.

Laboratory instruments are sold with the same rationale. If logic controlled instrument sales, there would be no need for four-color brochures, only long printed lists of specifications and dimensions. It is important to listen to what the salesperson is saying when you are buying instruments. A good salesperson asks for a buying decision. He or she asks you what you like best about the instrument or systems you now own while guiding you to make a decision to buy. Good selling is not about the killer sales close; it is about helping customers find out what they like about the system and showing them what you have that fits their needs or wants. A good salesperson never presents a feature *(logical)* of the instrument without describing its benefit *(emotional)*. Feature/benefit selling and guiding questions (ears) are the keys leading to a decision to buy.

Is this manipulative? It can be, but it does not have to be. I want to provide customers with what I feel is the best fit for their analytical needs. I want to find the best fit for their budget and for the problems they must solve. Because the decisions customers make appear to arise from their subconscious, I appeal to them emotionally through their feelings and their senses of touch, smell, and hearing. If I had access to customers' sense of taste, I would use that as well. But if I use their eyes in making the decision, it will be only through colors and pictures. These are the tools that allow the customer to feel the benefits of the instrument in solving problems and in accepting my ability to make the customer successful.

10.2 REASONS FOR AN INSTRUMENT SELECTION

Much of scientific research involves *deductive reasoning*. We take things that we want to understand and separate them into their component parts. Once we have done this, we try to analyze and understand the components' structure and functions. The usual next step is to use *synthetic reasoning* to reassemble the components into their original form to understand how and why they go together in that particular way. An alternative goal is to understand how to synthesize similar compounds that would fit in the organization to create new and unusual analogs with new and different properties.

Scientific instruments are used to analyze, separate, and simplify organized mixtures and compounds. They are used to tear apart the original structures, isolate their components, and study them. Instruments are then used to recombine the components into their original form or into new compounds.

Research is described as standing on the shoulders of giants in order to find new directions in which to proceed. If we want to find an instrument to separate a compound or mixture, we review the literature to see how other investigators have made similar separations. We find the instrument they used and seek the same or a similar instrument. We contact the manufacturer of that instrument or the competitors to determine the cost range for these instruments and the variables that these instruments can control or affect.

With this information, the next step is to find the money. A research proposal is assembled and submitted to a funding organization. When approval is obtained, we use the same information to prepare a bid proposal to submit to manufacturers of the required equipment.

Often we make a decision to buy in order to keep up with other prestigious investigaters. Another approach to obtaining the correct equipment is to find out what other researchers in the company are using and who they purchased it from; based on this information, we prepare the funding and bidding proposals. Still another approach is to talk to other researchers in the field at technical meetings, by e-mail, or by phone to finding what they are using and order the same instruments.

Scientific research is a very competitive undertaking, and new instruments can provide an edge if correctly selected. Instrument manufacturers attempt to continuously improve their products to give you that competitive edge, and they promote these features in their brochures and literature, showing their potential benefits for your research. Knowing this to be true, a potential buyer will look for the *best and latest* improvements in the form of increased sensitivity, increased sample capacity, increased speed of operation, and decreased separation or analysis times.

If the buyer has trouble determining these features, the manufacturer's sales representative will be happy to describe them. It is the buyer's job to determine if these features will be useful for his or her analysis and whether they are worth the increased cost. The buyer must also decide if the instrument is flexible enough to remain useful if the research requirements change as new directions appear in the separations, as they often do. These new features may be critical in allowing the investigation to proceed toward analysis of tinier amounts of more highly purified material.

10.3 PURPOSE OF THE DECISION

The longer a researcher has worked in a field, the more sophisticated he or she becomes in selecting research instruments. This translates into a better understanding of the variables needed for the instruments and greater ability to sort through the features and benefits that are needed. Generally, this means getting more instrument for your research dollar.

New investigators often lack this understanding and are often sold more capability than they need at a higher instrument price. Since they often have a limited budget, they may buy less instrument than they need from a company with limited field service and little if any application support. It is very easy for all but the simplest instruments to turn into a research project rather than a research solution. It is easy for a research laboratory to struggle to get an analytical instrument up and running and lose focus on the purpose of the project.

Instruments for a cost-per-test laboratory are usually purchased for a dedicated analysis, and can often be much simpler and less expensive. The exception to this is a methods development instrument with the flexibility of a research instrument.

The usual cost-per-test instrument must be rugged because it may be run on three shifts a day to generate the maximum number of tests. It will be used to generate data for forms generation to support customer billing, so additional computer equipment may be needed to combine the instrument data with other analytical and sample information, which will increase the system's cost.

10.4 PATH TO A SALES DECISION

The purpose of a professional sales presentation is twofold: to help customers decide what they want to buy and to help them find out how to buy it. Most people believe that when they decide to buy an instrument, they already know what they need. This may be true, but a win/win salesperson's job is to understand that often people do not really know what they need. The salesperson's job is ask enough questions to find out what customers truly wants, and then guide them to understand what they really need and show them how to get it. Usually at this point there is very little need for a major selling close.

A nonprofessional salesperson tries to force customers to buy what the salesperson wants them to buy, whether they need it or not. Because customers sense that they are being steered toward something that may not solve their problem, they resist the sales process and the salesperson feels the need to use manipulative power closes. He or she tries to force customers to buy *for their own good* because the salesperson has failed to do the necessary homework and find the best fit for the customer's needs. An example of such a power close is one called "Back the hearse up to the door and let them smell the flowers," used by the insurance industry in the 1930s.

The salesman in this scenario is trying to sell life insurance to a young father and mother when a child, who is being put reluctantly to bed, interrupts them. The salesman says to the father, "That's a beautiful child you have there, sir. She reminds me of a little girl I saw quiet recently. I was trying to help her family get life insurance when we were interrupted, just like we are tonight. Because of the interruption, we never got the policy signed. Last week, I opened the newspaper and I almost cried. That young couple was killed in a car accident, and I think all the time about that poor little girl. Why don't you go ahead and get this policy signed right now so that we don't have to worry about your little girl?"

That is manipulative power closing at its worst. It is not the action a professional salesperson should consider if he or she has the best interests of the customer in mind. It occurs because the salesperson failed to do the job. The customer in this example may have needed the insurance, but the salesman failed to find out what he needed and make the sale based on the benefits of his product; instead, he had to resort to a naked emotional appeal.

There is obviously a better way. A true sale is built on questions of two types. *Open questions* are used to draw out the customer's interests and needs. They are the classical "what? where? why? who? how?" questions that reporters use to elicit information. Rather than argue with a response, a salesperson will ask another open question or will respond with "Oh?" in order to get the customer to rethink or clarify his or her response.

The second type of question is the *closed question*, used to guide the customer to the decision. Closed questions usually can be answered only with "Yes" or "No"—for instance, "You did say that you wanted the instrument in blue, did you not?" Often a closed question is a compound statement made with a closed question at the end: "This gradient system will really solve your research needs for the near future, don't you agree?" Another closed question that you will hear salesmen use is "Is that fair?" Everyone wants to be fair, and customers will often reply affirmatively to this question. If the process is combined with the knowledge gained by the salesperson using open questions in the information-gathering process, it can lead, step by step, to an emotional decision to buy based on the logic of the customer-supplied information.

10.5 THE QUALIFYING SALES INTERVIEW (ADMANO)

As a new salesman for an instrument manufacturer, I talked with the most successful salesperson in the company at a sales meeting. I asked him, "What information do you seek when you first approach a new potential customer?" I was told that he wanted to know DAMAT: who the *decision maker* was, what the *application* was, how much *money* was available for instrumentation, when it was *available*, and the *timing* to finish the order.

I have been told that genius is seeing something that everyone has seen and thinking things that no one else has thought. It is important to take material you are given and customize it to fit your needs. I took the best salesman's DAMAT idea and modified it based on personal experience into a system that has served me throughout my sales career.

The information I sought on a first interview was the *application, decision maker, money, availability, next step*, and *others* involved with the purchase—ADMANO. This can be done either over the telephone, by e-mail, or face-to-face at the customer's institution. I sought the *application* information immediately because I could often get it from a third party—for instance, a graduate student or a colleague—even if I had not yet found the decision maker. The *decision maker* is next in importance; it is usually a waste of time to make a presentation to someone who cannot buy.

Once I know what the application is and who will make the decision, I need to know how much *money* is available for the purchase. Surprising, when I have established trust and explained why I need to know how much money we have to work with, most of my customers are not reluctant to disclose their instrument budget. I also need to know how soon the money will be *available*. Grants, fellowships, and awards almost always have release timing, and a purchase cannot be made before this time. The customer can put together system requirements and proposals beforehand, but a purchase order cannot be issued until the money is available.

Once the money issues are dealt with, the customer and I both need to verify the *next step* in the sales process. One should never leave a customer without knowing what needs to be done next. Many sales founder because the next meeting had not been set up. Without this clarification, things drift and get lost, including the final purchase order and installation. Do not put in the effort and time if you are not prepared to go all the way to a completed purchase.

The final issue is to determine what *other* persons are involved in the purchase decision. These persons consist primarily of students and other researchers involved in the application for which the instrument is being purchased; gurus, bosses, and colleagues who advise and consult in the buying process; and approvers of the money release.

Generally, there is someone in the company whom the principal investigator (PI) goes to for advice about the instrument selection. This may be a department chair, a senior graduate student, a postdoctoral student, or another investigator doing similar research. If you fail to find this person early in the interview process, he or she will often turn out to be an adversary blocking the sale later on. Find these persons early and convert them; they can provide invaluable assistance in pushing the purchase through the purchasing bureaucracy.

I have already mentioned that students and technicians often do the research and recommend the actual system that is purchased by a PI. The PI in most places is usually so busy guiding research, writing grants for new sources of income, and reporting to higher authorities that he or she often delegates much of the purchase decision to others. Many instruments are so expensive that they have become multiuser core instruments whose cost is shared by the department or other investigators. All of these people must be consulted during the planning process to determine who the decision makers are. Generally, the one with the most money involved dictates many of the rules controlling the purchase.

Although PIs seek grant money for their own research, they are not completely autonomous in making the decision to spend it. New investigators may have received departmental or university startup money in setting up a laboratory, but its use must be approved by the source. Even federal grant money and funds allocated for company projects require approval and reporting to financial administrators. Most universities require grant money to be submitted to a financial accounting department that extracts a fee from each grant.

The sales interview is a very valuable tool. I did not always use it early in my career. Sometimes my initial telephone call and visits lacked sufficient time to complete the process. It is important to keep notes on where you are in the process and add information as you proceed. But the sale was always smoother and more successful when all the information from the ADMANO process was completed.

11

PROFILING THE SALES CALL

This chapter was written to explain what should take place in a sales call, not to encourage you to become a sales representative. The world of selling is already overcrowded with bad salespeople who have given the selling field a terrible reputation. The top 10% of professional salespeople make 90% of the sales and 90% of the income in the field, as in most professions. Professionals in any field become professional by studying and practicing until they improve. Where possible, you, the customer, should try to work with professional salepeople in making your purchases. It will save you time, help control your blood pressure, and ensure that you obtain the equipment you need.

11.1 TRAINING SALESPEOPLE

Bad salespeople are fairly easy to identify by their clumsiness; you probably will recognize them from your previous experience with direct marketing telephone calls and used-car sales presentations. They sound like a poorly written list of specifications for the instrument. They fail to use questions to find out exactly what you want and need from the product they are selling, They ask for the order without earning the right to close. And last of all, they won't stop talking and listen.

Buying and Selling Laboratory Instruments: A Practical Consulting Guide.
By Marvin C. McMaster

Typical of this type of salesperson is Fast Freddy. Freddy started out in direct marketing sales and moved up to instrument sales. He began by selling milking machines, reaching the height of his career when he sold two milking machines to a farmer with only one cow and took the cow in down payment—an example of a win/lose sale that benefited the salesman but not the customer.

Most people get into sales by accident, economic necessity, or failure to earn a living in any other field. Used-car salespeople are said to have a turnover rate of 110% a year. This occurs because few organizations take the time to train new salespeople to become sales professionals. They give them brochures, a five-minute talk on selling, and a quota to meet; then they throw them into the water to see if anything gets bought.

A very few salespeople buy paperback books on selling and personal motivation, read them (I read 72 selling and personal development books in my first year as a salesperson), start practicing, and eventually get better. A very few of them last long enough to develop professional pride and come to understand that they succeed only when the customer succeeds. These people learn that they must be a partner in the sale, as well as a consultant who helps customers find and understand what they need to buy and shows them how to buy it. They serve the customer and in the process become true salespersons.

My sales training experience was a little unusual. I had received technical training before I began to sell and had used the types of instruments I was going to sell in the laboratory. I had little sales training before I started. The instrument company provided me with an unusually complete background. They sent me to the manufacturer's headquarter for a month of training. They provided a binder full of brochures and specification sheets, hands-on training on the instruments I was to sell, and a generic course on selling bulldozers. I was taught, and practiced in seminars, to use listening skills and to ask open and closed questions to draw out customers and guide them to a buying decision. Finally, I was allowed to shadow a successful salesman as he made telephone appointments and sales calls for a short period of time.

I was given a geographical territory, a quota, a list of prospects for my company's equipment that had come from responses to technical magazine advertisements, and a list of current users of our equipment in major accounts in the territory. I went into existing accounts as a new salesman, and because of my laboratory experience, I entered as a technical colleague. I asked customers what they liked and disliked about the instrument our company sold. Where possible, I brought ideas that I had picked up from my experience and from the company's training. I decided to become a consultant bringing ideas into the laboratory and taking away ideas that I had learned from these experts when possible without violating company confidentiality. I began to organize this body of information into selling systems.

11.2 HOT BUTTON ANALYSIS (HBA)

The single most valuable tool I was given as a salesman came out of a two-hour advertisement for a course called "Managing Interpersonal Relationships" offered by Wilson Learning of Minneapolis. I encountered it while taking a week-long selling course offered by the same company called "Counselor Selling." As I worked with the ideas offered in the two-hour teaser presentation, it evolved into a technique that I have come to call *hot button analysis* (HBA). I later found that, as usual, we stand on the shoulders of giants. The original technique was invented some 3000 years ago by Pythagoras, who had developed the theorem for calculating the length of the sides of right-angle triangles. Pythagorus believed that all people fall into four categories and that different things motivate people in each category.

11.2.1 Verbal/Visual Placement

The key to modern use of this technique is to recognize that simply trying to place a person in a category by recognizing his or her "type" is generally ineffective. Classification of types is much more accurate using a verbal and a visual guide for placement (see Figure 11.1).

Figure 11.1 shows the use of verbal coordinates on the X-axis and visual coordinates on the Y-axis. On first encountering an individual, ask yourself, does he or she tend to ask or tell when speaking to you? Someone who asks obviously uses a lot of questions; someone who tells makes statements. Next, ask yourself whether this person masks or shows emotions. Someone with a good poker face and unreadable eyes tends to mask emotions; if the eyes and face are animated in conversation, emotions are displayed. A person who tells and masks is called a Driver. Telling and showing emotion places a person in the Expressive category. Someone who asks and shows emotion is an Amiable. Finally, a person who asks and masks is called an Analytical.

11.2.2 Handling Graded Placement

Obviously, there are people who do not completely fit into the categories shown above. They may, for example, have some of the characteristics

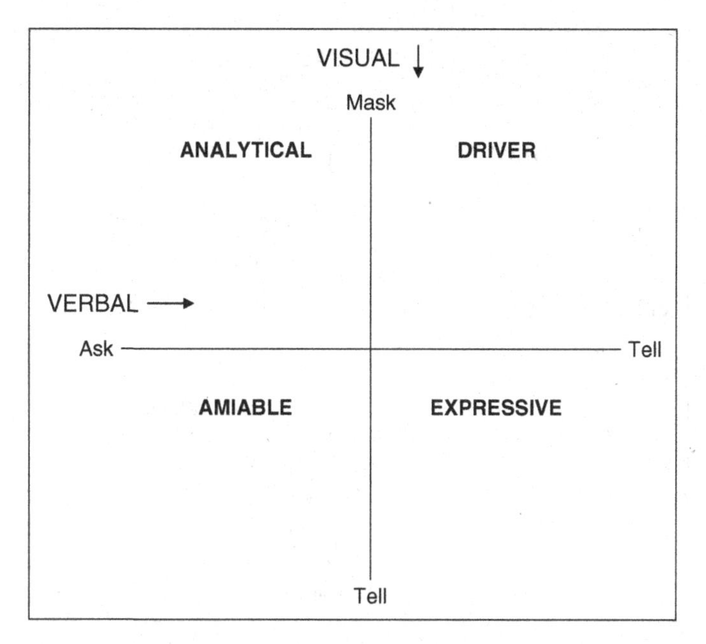

FIGURE 11.1 HBA Verbal/Visual Placement Diagram

of both a Driver and an Analytical. To provide for these individuals, we have to add gradations to our two axes (see Figure 11.2).

The X-axis is labeled with letters, the Y-axis with numbers. A person who tells most of the time but always masks would be labeled an A2 Driver. Someone who always asks questions and shows emotions is considered a D4 Amiable. An individual who is very hard to read, who shows some emotion and tells some of the time, might be classified as a C3 Amiable. Obviously, the model begins to fall apart in distinguishing these mixed-category individuals; try not to get too involved in subdividing classifications.

The next step is to label the diagram with hot buttons that apply to each category (see Figure 11.3). A Driver's buttons are in the area of *control*. She wants to control people, time, money—all the situations of her life.

The Expressive is motivated by *recognition*. He wants to be first, the most important, and he wants to win the Nobel Prize so that everyone will know that he is important.

The Amiable wants *security*. She wants to sleep soundly at night, knowing that all is well at the office or laboratory. She tends to buy automated equipment that can be run unattended. She wants to trust you, and if there are any problems, she wants to hear them from you.

The Analytical wants *peer respect* or *status*. He wants details and specifications, the more the better. He wants to make sure that he is making

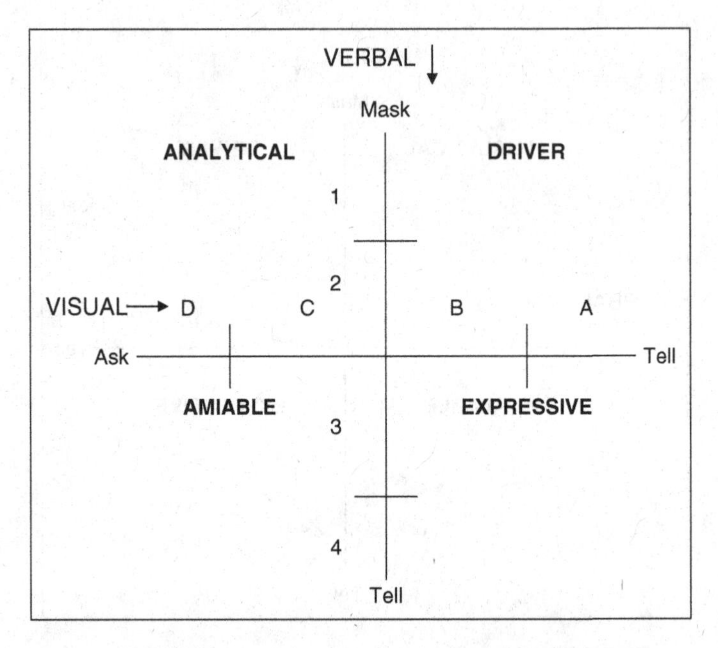

FIGURE 11.2 HBA Axis with Gradations Added

the right decision. If he wins the Nobel Prize, he can be sure that all of his friends know who he is and respect him.

11.2.3 Motivational Hot Buttons

These hot buttons for each type are a communication tool that will help you in all your dealings with people if you choose to use it. The HBA technique takes about 30 seconds to apply. It lets you target the person before you. It is not limited to use by salespeople. My wife is a home care registered nurse. She is a better user of HBA than I am because she always remembers to use it and always uses the verbal/visual axis. The two-hour sales presentation I attended that explained the tool was designed for technical service representatives so that they could better serve their customers.

One word of warning: this tool is often ineffective when you try to use it for self-analysis. It is hard to make an objective analysis of your own verbal/visual styles. Ask your friends or your spouse to help. Explain the tool and have them apply it to you. Others classify me as an Expressive A4, although I initially thought I was an Analytical or a Driver.

My wife assures me that I am an Expressive. You may find that you are different types at different times and in different settings. At home in my rest mode, I tend to move more toward the Amiable side of my nature.

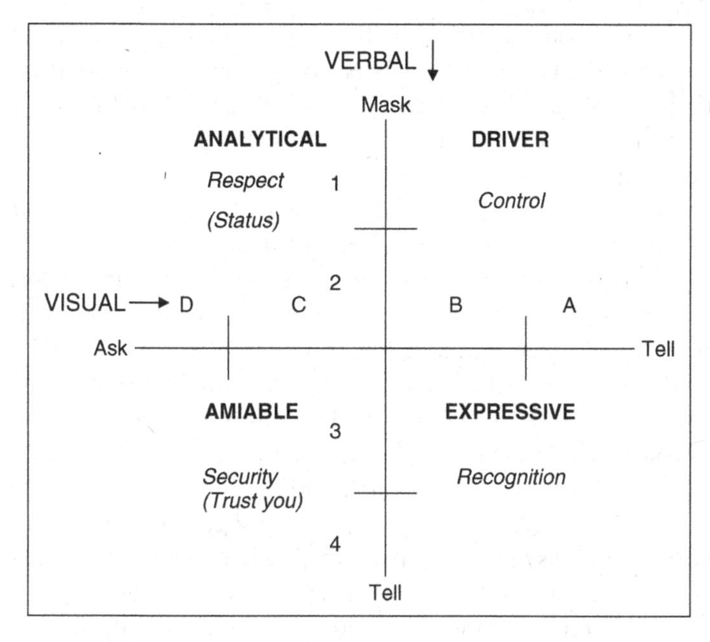

FIGURE 11.3 HBA Motivational Chart with Motivations

11.3 SELLING TO EACH HOT BUTTON TYPE

When you are dealing with a Driver, it is useful to match his or her style; offer control choices that are win/win for both of you. This tool is obviously not limited to use by salespeople. Drivers do not like people who waste their time, and a salesperson who does so seldom gets a second appointment. When I recognize a Driver, I say, "I would like to come back on Tuesday at 2:30 P.M. with two proposals I have prepared for you. One is a bare-bones configuration; the other has all the bells and whistles. Each will be labeled with its price, so you can decide which one you want. To prepare the proposals, I will need to ask you five questions. Is that fair?"

Drivers almost always say "Yes." I ask them ADMANO questions and leave. When I return, they make their choice, the order is placed, and the decision is final. Most salespersons have a standard presentation for all types of customer. A Driver will listen for about three minutes and throw them out of his office. They both lose.

Each type has an individual approach. Expressives relate to success stories. They want to see the big picture and are turned off by details; things slip through the cracks. They definitely live in the moment but will plan for the future if the process does not have to be too detailed. Expressives need a notational calendar and a priority-based schedule to get things done. You do not need to get to the point with Expressives.

Tell them a story on the way out the door and they are ready to buy. Expressives are fickle. They will listen to your story, decide to buy from you, and change their minds when the next salesperson with a story shows up. They require follow-up until the instrument is in the laboratory and up and running.

Amiables are warm puppies; they want to be your friend. But make sure that they hear any bad news about your equipment from you. Once you lose their trust, it is very difficult to regain it. Be sure to locate and talk to the Amiable's guru if there is a problem about the order. If the guru is also an Amiable, losing his or her trust can affect the whole account, whether it is a company or a university. The Amiable's office is almost a stereotype. The desk is always on the wall across from the door and is often turned sideways to the door. Just inside the office, right beside the door, is often a comfortable couch and a table with a coffee pot. The Amiable will get up from the desk, come to greet you, ask if you want a cup of coffee (the answer is always "Yes," whether you drink it or not!), and then sit down for a talk to establish that you are trustworthy.

Do not rush into an immediate sales presentation. Talk instead about guns, dogs, and kids. Ask about his or her family. Listen to this person's needs and problems. Demonstrate how your automated equipment will provide solutions and security. Drivers and Amiables get along fabulously, although the Amiable is probably very uncomfortable in the Driver's office because the Driver is forted up behind a massive desk separating the Driver and the Amiable. You will not have to worry about Amiables after they have placed the order. You are their friend, and they take care of their friends.

The Analytical will drive an Expressive sales representative totally crazy. Analyticals can never get enough brochures, specifications, and details to make a decision in one visit. They always second-guess their decisions. When they finally make a decision and issue a purchase order, they continue to worry about it. Stay in touch and do not be surprised if they change their minds. Analyticals will probably not buy on the first sales call. They will have to be cultivated with more and more information.

Formerly, I had to use a negative close to get an Analytical to buy: "People would laugh if they heard you did not have this instrument in your laboratory." I used it, but I felt it was pressure selling. Eventually, I found a better way. I would offer the Analyticals a variety of features, and they would argue with one of them. I would wait until, on a later call, they offered me the same feature as their idea and defended its importance. I would agree with that feature's importance and close the order on the basis of that feature. If the Analytical started to waver after the decision, I would restate how important that feature was to him.

Analyticals have to internalize the idea for themselves to own it. Let them own it! Do not remind them that you told them about the feature in the first place. Egotists will lose this order by debating who had the idea first.

So, there you have four types of people with degrees of shading. Their hot buttons are usually their major reason to decide, but they are motivated by other hot buttons as well. The biggest problem is remembering to use the verbal/visual tools to place people in their correct boxes before estimating their hot buttons. As you become more familiar with HBA, you may tend to try to skip the reading and just place people in the boxes. Your success rate will fall rapidly, and you will wonder what happened. When I take my verbal/visual readings at the first call, I am usually about 80% accurate. If I just try to place people in a box and estimate their hot button, my accuracy is less than 50% when I later review the call.

No tool works perfectly every time you use it. The intermediate placements in the B and C and the 2 and 3 categories will have you scratching your head, and your accuracy will suffer. However, this is better than no tool at all. You will still be in the ballpark even if you come up with combination hot buttons.

If you have run the visual/verbal coordinates and come up with a hot button, try matching the characteristics when you talk to customers. For example, if you are an Expressive working with a Driver, match his or her speech and rhythm but let the Driver have control. Guide this person with your knowledge. The process is more of a pull than a push. You will gain technique as you begin to apply the tool.

Try using this technique with your colleagues in an administrative meeting. Get into the habit of using it in every meeting you attend and begin to form the habit of finding hot buttons for everyone you talk to. You will find people easier to understand and work with when you understand their hot buttons.

Practice, practice, practice.

12

OBJECTIONS IN THE SALES PROCESS

The great sales trainer Zig Zigler says that there are five objections that need to be overcome in making a sale: No trust, No need, No help, No hurry, and No money. Most people would consider the last to be the major objection, but this is usually not true. The money almost always exists, but the purchase may not have a high enough priority in the customer's mind to justify the expense. The salesperson's job is to increase the value of the product and the urgency of the customer's need for it.

12.1 SYSTEMATIC SELLING

New sales representatives have an irrational fear of objections due to the belief that the customer is objecting to or rejecting them. Objections are a sign that customers are interested! If they do not object, they probably do not care and will not buy unless further selling is done. People who are interested in something and buy it still find things about the item that they do not like. Empathy is very important in handling objections since selling is done emotionally.

If you sympathize with an objection, you will feel the same way as the customer and will be unable to handle the objection. If you empathize with the customer, you understand how the customer feels, but you do not

Buying and Selling Laboratory Instruments: A Practical Consulting Guide.
By Marvin C. McMaster

feel the same way and you can therefore be part of the solution. Generally, objections grow out of misunderstanding. By clarifying the problem, you can usually come up with a way of handling it. The salesperson knows more about the instrument than the customer and can use that information to handle most of the customer's objections.

One classical way of handling objections is called *feel, felt, found*. It is basically a way of applying empathy. The salesperson says, "I think I know how you feel. I felt that way too, until I found out that ...." At this point, you can introduce your solution, assuming, of course, that you have one that will correct the customer's misunderstanding.

Use subtlety in applying this technique. A customer who has ever bought a new or used car has heard "feel, felt, found" many times and immediately treats you like a used-car salesperson. A customer once told me that my gradient HPLC system was far too expensive. I could have discounted, sold down to an isocratic system with only one pump, or moved to what I felt was an inferior gradient system that would not serve his needs. Instead I said, "When I first heard how much a system like this cost, I too was shocked by the price, but after I talked to some colleagues at a meeting about what it could do, I realized how much research time I could save by using it and decided it was worth every cent. Your time is money, isn't it? My boss tells me that all the time." This was a simple use of "feel, felt, found," but it worked to get the order that time.

The teachers of the course on counselor selling that I attended at Wilson Learning agrees with Mr. Zigler that there are four objections that must be overcome in order to make a sale (see Figure 12.1)

Let us talk briefly about the *No money* objection. The customer is running a research laboratory in a university or in a company. He has to have a budget to do his work; if not, he is out of business. What he is really saying is that he has the money, but your equipment is not a priority on his equipment funds list. That's fine. Your job is to make it a priority

	Objection	Solution
1.	No trust	Counselor credentials
2.	No need	Feel the fever, create desire
3.	Ho help	Features and benefits
4.	No hurry	LSCPA

Additionally:

Objection	Solution
No money	Increase the priority
No sale	Get permission to sell again

FIGURE 12.1 Handling Objections as a Counselor

for him. That is called selling. There is always money; it just depends on what he is going to buy and what he is going to give up for the moment. The only question is, what will he be buying today from someone? Let's make it you, shall we?

12.1.1 Establishing Trust

First, you must overcome the problem of *No trust*. The customer who does not trust you will not buy from you. If customers don't know you, why should they trust you? Other salespersons have sold to them, and they don't trust them. When you walk in, you look like any other salesperson. Maybe they have heard of your company from colleagues, which may slightly lower the trust barrier. A business card and a shiny brochure introducing the company and its products will help to validate your employer.

The way to overcome the trust barrier is full disclosure. When I walk into a prospect's office I do a fast HBA so that I can understand what motivates this potential customer. I introduce myself, my company, and my background in using the equipment; then I state how I came to hear about his or her interest in my equipment. From my ADMANO interview, I find out what the prospect wants to do and how much money is available for the purchase. I then explain that my job is to act as a consultant for this person in the buying process.

I have told my customers, "I am a rather unusual salesman. I am a professional with a Ph.D. in organic chemistry and postdoctoral degrees in in biochemistry and I have audited medical courses. I worked in a process laboratory solving plant problems with an HPLC similar to the one you are considering. Because of my degree, I am rather an expensive employee for my company, and to justify the cost, I have to sell four systems instead of one. If you buy only a detector today, eventually you will expand that to a system and eventually you will help me sell four systems. Now you may not buy all four systems personally; you may simply refer me to your friends, who also will buy. You might ask, why would you do that? Because my job as a consultant is to make you so successful that you will want to buy another successful system and refer me to other people who would also like to be successful. Is that fair?"

Did that really work? Yes, it did, repeatedly, because it is the truth. I can refer you to customers who have bought equipment from me in my work with four different companies over 25 years. They bought knowing that if they had a problem, even with equipment from one of my former companies, I would intervene on their behalf and help them solve their problem because I take care of my customers. Because of this, my customers have helped me to become a million-dollar-a-year salesman.

The people I sold to bought expensive equipment. They wanted help in buying it and making sure it worked to solve their research problems. It is hard to find a salesperson who cares about them and about their work as much as they do. The consultant approach will work just as well with equipment of any price as long as it is used honestly. People buy from salespersons they trust. If that trust is violated, they do not buy again. Much of my problem in sales consisted of cleaning up messes created by other salespeople. Zig Zigler says that we don't have a recession; we have a shortage of good salespersons and a surplus of dishonest business persons. Your own experience as a customer will confirm this statement.

12.1.2 Helping to Create Desire or Need

The next objection that needs to be handled in the sale is *No need*. If a customer trust you but does not feel that your equipment is needed, he or she will not buy. Someone with a high fever who does not feel hot will not take something to reduce the fever. Potentially, this could mean killing oneself through inaction. My job as a salesperson was to help the customer feel the fever and take action.

Creating desire requires information. The process begins with open questions of the form "who? what? where? when? and how?" to gather information on the customer's interests and applications. Once it is clear where the customer need help, the salesperson can use close questions that have "Yes" and "No" answers to help the customer understand what should be purchased and to help the customer develop enthusiasm for the idea of solving the laboratory's research problems. Sales tools such as technical articles, brochures, and specific referrals to successful customers can all be used to help achieve this goal.

Often a potential customer has already sought out information on a product. This is the tire-kicking stage of a sale, usually in response to interest raised by some type of advertisement the customer or laboratory personnel have seen about your product or a similar product from a competitor. In a technical sale, I usually get involved when a customer responds to an advertisement in a technical journal or sends a letter to the company expressing interest in a product he or she had seen in a journal article or heard described at a technical meeting or by a colleague. This person has some interest in the product and may know that it is needed to solve a problem, but is not motivated to purchase anything right now. He or she does not feel the fever.

A customer's motivation to buy equipment is to accelerate the research and the publications that emerge from it. Your instrument will effectively be a time machine to achieve his or her research goals. An ADMANO interview provides information on the customer's research application and whether he or she has money or will need to acquire it for a purchase. The open questions about the customer's applications interests will reveal information about what this person needs to do to finish the work. Usually help is needed to increase the sample load, to speed up the analysis, or to increase the sensitivity of the analysis. When a potential customer talks about major application interests, he or she will begin to become excited and involved.

Once the customer had provided the major needs of the work—say, speed of analysis or sensitivity in detection—the salesperson will move to closed questions to guide the customer's thinking about how the salesperson's equipment can provide a major increase in speed or sensitivity: "I think you can see that our system is a time machine, and you must realize how important your research time is to your success, don't you?" For instance, a typical closed question might be, "Have you thought about how a tenfold improvement in separation between the protein you are after from its impurities would increase the quantity of material you could inject and the amount of purified compound you would recover? That would give you a real competitive edge in getting your results and papers out faster, wouldn't it?"

Time is money and results to the potential customer. It balances out the time and effort needed to file for a grant to purchase the equipment. It balances out the time needed to train laboratory personnel. It offers the potential to handle other research problems when the current problem has been solved.

At this point, the customer trusts you. He or she understands the benefit of your type of equipment; can see the laboratory people using it to solve research problems; and wants to have it in the laboratory. But he or she is not yet convinced that it will actually work or that the laboratory people can learn to use it on a regular basis. The customer needs proof that it is worth the time, cost, and effort that will be needed to obtain it.

12.1.3 Showing That Your Equipment Will Help

The next step is to handle the objection of *No help*. At this point, the customer trusts you and understands that the equipment is necessary. Now the customer must be convinced that the equipment you are offering will actually do what you claim it will do. The proof statement of your claim must come from brochures with their pictures, examples, and specifications. Articles in the literature that mention using your equipment and referrals from current customers who are using it will further validate your claims. If the instrument you are offering is newly introduced, you may have to offer an in-laboratory demonstration of it actually running one of the customer's samples.

You may also have to distinguish your equipment from competitive instruments. You need to highlight not only the *features* of the equipment, but also how these features will translate into benefits to the potential customer's specific research needs. Where possible, you should emphasize features and benefits that are specific to your equipment alone and show how they will meet the customers needs.

12.1.4 Creating Sales Urgency

The last objection that must be overcome is *No hurry*. The customer trusts you enough to listen; realizes that he or she needs the type of equipment you are selling; and believes that your equipment is exactly what is required. But the customer feels no urgency to part with the money or to put in the time and effort right now to write up and submit a proposal for money. He or she is caught on the horns of a dilemma, wanting the instrument but not wanting to spend the money or the effort to get it. The salesperson's goal at this point is to close the sale and get a purchase order as soon as possible. This step may not even be necessary if the earlier steps of making the customer aware of a need and convincing the customer that your instrument is necessary have been done correctly. The customer will simply buy.

Wilson Learning's technique recommended for handling lack of urgency is a variation of "feel, felt, found" called *LSCPA*. That stands for *listen, share, clarify, propose a solution, and ask for the order*. You listen to the customer say something like "I sure would like to have your gradient HPLC system because I can see it would really speed our research, but I just earmarked a lot of money for our new spectrophotometer and my laboratory funds are exhausted until January. Come back then and we'll talk."

You realize that things can change a lot in six months. The customer's research needs this system now, and waiting six months simply delays the results and the publication date. You share the customer's feelings: "I know how you must be feeling. I remember what it was like when I was struggling to balance equipment money when I was in the laboratory." Next, you clarify the problem: "Don't you think the HPLC system might have more of an impact on your research right now than a new spectrophotometer? The old spectrophotometer is still working, isn't it? Why not get the gradient system in right away and postpone the new spectrophotometer to the next grant?' Then you propose a solution: "Look, I can get you shipment by the end of the next week. I know there will be a pricing difference between the two instruments. If I get you a ten percent discount in the form of solvents, columns, and accessories from our catalog, you will have everything you need to get up and running fast. As soon as the system is in, I'll come in with my service man to help you get your application up and running." Finally, you ask for the order: "Could you get a purchase order cut by tomorrow so that we can get going?"

Six months from now, you will have to start selling all over again. Do not waste your selling time; get the order moving today. The worst that can happen is that the customer will say "No" and leave you where you are now.

If the money difference is too great, I might propose a different solution: "An isocratic single-pump system would be less expensive that the spectrophotometer. Why don't you get an isocratic system now? I'll get you column, solvents, and syringes at a ten percent discount and have our application laboratory work out a separation for your samples. You can get up and running and get your users familiar with your HPLC separation. In January you can budget for another pump, controller, and mixer, and I'll help you reconfigure it into a gradient system. Let's go ahead and get a purchase order cut for the isocratic system today so that I can get it and the supplies in here next week, Does that sound fair?"

If I wait until January, many things could happened to the sale. The competition may hear about the order and show up, the customer will have lost the use of the HPLC for six months, and the whole sales process will have to be repeated. By placing the order now, the customer gains free solvents, columns, and supplies worth 10% of the cost of the instrument, free support from the application laboratory, and my time and effort to get his system up and running on his research application. I get the order now, in a competitive sales environment, and will have a successful customer as a positive referral in my account; both the customer and I win in this situation.

12.2 ASSISTANCE OF SALES TOOLS

Sales tools will not make the sale, but they help validate the claims that the salesperson makes. Brochures with pictures of the equipment that is being offered prove that the equipment and the company exist and provide details, operational descriptions, illustrations, and specifications that a customer can use to see if the equipment will be useful. The salesperson's business card helps to establish trust and availability of people who can help provide needed information and equipment.

Referral lists and relevant technical articles provide examples of people who have used the equipment to run similar experiments. They help to prove the customer's need and demonstrate that the equipment will help to quickly solve the customer's research problems. Often these materials are sufficient to convince the customer to buy your system, especially if the equipment represents a well-documented, well-developed technique.

Sales seminars and poster exhibits at technical meetings provide technical details of instruments and their application to various research projects. They can be as general as an overview of the theory of an instrument's operation or as specific as details of the instrument's use on a single application. They are useful to a salesperson for creating interest in the instrument and are usually used early in the sales process to create a list of potential buyers.

12.3 USE OF DEMONSTRATION EQUIPMENT

Equipment for use in newly emerging technology will often have to be demonstrated to convince customers that it will solve their problems. There are two basic types of instrument demonstrations.

The dropoff demonstration is used only with a customer who is familiar with this type of instrument and is considering a newer or more advanced system. Hopefully, it will come with an installation either by a service representative or by the salesperson. Always remember that Murphy's law always accompanies demonstrations and be prepared to respond to cries for help. Once the equipment is up and running, the objective of the sale is to keep this equipment in the laboratory permanently by offering it as a demonstration instrument. The sales representative usually requests this before bringing in the equipment: "If the equipment works as advertised on your project, would you be able to purchase it right away as a demonstration unit before I have to pay to ship it back?" The 10% demonstration discount is fairly standard in the industry, or an equivalent amount of operating supplies might be offered.

The salesperson brings the other type of demonstration system into the laboratory on an instrument cart. It is critically important to determine ahead of time the goal of the demonstration that will lead to a purchase. The salesperson must establish what the customer needs to see from the instrument in order to place an order for similar equipment.

As a young sale representative, I was asked for a demonstration of an HPLC system in a government account six hours from my home base. When I arrived at the laboratory, I found that the customer wanted to see separations of three different types of compounds, one of them at the maximum possible sensitivity. I was able to do all the separations, but I had to stay overnight and come back the next day. Later, I found out that the customer lacked funds, and had talked about buying this type of equipment for four years and never bought anything. He was just tire kicking. I could have avoided this problem by doing a much better job of asking qualifying questions before I set out to make the demonstration call.

The secret of an effective demonstration is to keep it as simple as possible and to leave time for closing when the customer is satisfied. I found that for a laboratory demonstration, the salesperson should know the separation to be achieved. Practice it before going in, if possible, or do a fast scouting separation to find the conditions and use this to set up an analytical separation.

I once had a customer who wanted to see our new autosampler demonstrated at its minimum advertised sample size. Since this was an established customer, I asked him if he thought it was reasonable to expect that level of performance from an unoptimized instrument. Finally, he agreed that it was not. I sold him the instrument with the agreement that if he would buy an autosampler and have it installed by a service representative, we would demonstrate the minimum sample size in our specifications or take the autosampler back and return his purchase price. Eventually, we showed him six 1 μL injections so consistent that I could lay a ruler across the peaks tops from the detector. I would never try that on a demonstration instrument that had been carried around strapped on a body cart in the back of a station wagon.

Just remember, objections are a sign of interest. The salesperson's job is not to handle objections, but to find out what equipment customers are interested in and to help them buy it.

13

STEP-BY-STEP INSTRUMENT SELLING

This purchasing situation is similar to the sample in Chapter 2, but seen from the viewpoint of the salesperson. The individuals and locations are purely fictitious.

Successful purchasing requires a knowledgeable buyer who understands the selling process. As a first approach to understanding what happens in an instrument sale, we will walk through a typical sale. It is not representative of all instrument sales, because I believe in a win/win sales process and not all salespeople do.

Mitch Baker is a salesman with Consolidated Analytical Instruments, a manufacturer of various laboratory systems used to solve biochemical problems. He has just finished a visit to new customers in a Western Indiana University Biochemistry Department to check on their satisfaction with their new HPLC system. Everything is going well, and they are excited about the results produced by their separations. Mitch asks the postdoctoral student in charge of the equipment, Sam O'Conner, if he knows anyone else in the department who is looking for new equipment. Sam mentions a new associate professor, Dr. Henry Jones, who is looking for a similar system to separate renal proteins and peptides. Mitch obtains Sam's permission to use his name when he sees Dr. Jones.

Dr. Jones invites Mitch into his office and says that he wants to talk to Mitch about some brochures that he had requested and received from Consolidated. However, the department chairman has just called a staff meeting. He wonders whether Mitch would mind talking to his

Buying and Selling Laboratory Instruments: A Practical Consulting Guide.
By Marvin C. McMaster

postdoctoral student, Dr. Tom Alonzo, who is spearheading his search for equipment and is more familiar with exactly what they will need; Mitch agrees.

Mitch asks how much money he has received for startup funding, what other equipment he will need, whether the money is immediately available, how soon the equipment is needed in his laboratory, and who else Dr. Jones will be consulting about the decision. Dr. Jones says that he has been given $150,000 for equipment, supplies, and salaries and has brought another $40,000 in grants with him from another Indiana University. The money is immediately available, but the system must to go out on university bids, and he will consult with the department chairman and probably with the laboratory personnel before making a decision; now he needs to leave. "Talk to Dr. Alonzo and work out the details." Mitch makes a note to himself that Dr. Jones tends to give orders and control his emotions. Dr. Jones is probably an A-2 Driver and will need options to choose among in making his decision.

Mitch then introduces himself to Dr. Alonzo, explaining that Dr. Jones has sent him over to discuss separation systems. He has just come from the laboratory, and Sam O' Connor has told him that Dr. Alonzo is trying to purify renal proteins and asks if that is correct. Dr. Alonzo explains that they thought the proteins were precursors for a neuropeptide hormone that is their true target molecule, so actually, they are trying to purify both in a large enough concentration to do animal studies.

"Well, maybe I can help cut through the process and the details for you," Mitch Baker says. "I did peptide synthesis with Dr. Demert in Florida and separated them with HPLC, so I am somewhat familiar with what you are trying to do. Have you got a bioassay set up for the neuropeptide?" Tom assures him that they do, and Mitch sits down and goes over the brochures, pointing out the advantages and drawbacks of all the systems they contain. It quickly becomes evident that a gradient HPLC system will be required to separate the peptides, but Mitch shows Dr. Alonzo how they could later separate the pumps to set up both a protein purification system and a peptide assay system by adding another injector and a simple detector when they finish their critical peptide purification. Mitch also shows him how he could do preparative purification on the same system by changing column types and pump heads. He helps Tom pull together a list of critical specifications for the bidding forms.

Mitch's HBA of Dr. Alonzo indicates that he asks questions and shows his emotions, which make him a D-4 Amiable. Mitch feels that Tom trusts him and would be interested in automated equipment to provide results security. Mitch knows he will have to make sure to tell Tom about any problems on delivery or installation.

Mitch asks if any other hardware will be needed to complete the separation. Dr. Alonzo replies that Dr. Jones would like to get a 2-D EP system to aid in separation and to confirm the purification, but they are not sure if they will have enough money. Finally, Mitch asks how soon they will need the equipment and who else will have to approve the final selection when the bids come back. Tom says that the sooner they get it the better, and that they are hoping to do a poster presentation at the FASEB meeting next year. He thinks Dr. Jones will be able to approve the bid, but he might need a nod from the department chair to finish the order.

Mitch says that he will have a pricing quotation for a gradient system sent to Dr. Jones. He mentions that Consolidated does not sell 2-D EP systems but usually recommends the one listed in the Federated catalog, and he volunteers to get them a discount on solvents, columns, filters, and syringes to help pay for the 2-D EP system if Consolidated gets the bid. He makes an appointment to go over the price quotations two days later for the HPLC and 2-D EP systems and to set up the lockout specification on the bidding forms that Tom will get so that they can get things underway. They shake hands and set up an appointment for 8:00 A.M. on Wednesday morning. Tom says that he will check with Dr. Jones to see if he can sit in with them while they finish the bid proposal and will call if there is a problem.

On Wednesday, Mitch appears in the laboratory. Tom and Dr. Jones write up the bid specification on Consolidated's HPLC and the 2-D EP system from Federated, and Tom leaves to hand carry the bid to the Purchasing Department. Dr. Jones spends a few more minutes discussing with Mitch other purchases he is considering after he gets additional grant money. Mitch asks if Dr. Jones knows anyone in the department or at his previous position who might be looking for additional equipment that Consolidated manufactures and sells. Dr. Jones gives Mitch the names and telephone numbers of two professors at the last University who are budgeting for HPLC systems, another professor at a college in Illinois who has told him that he is getting an HPLC system, and another new professor at the current school who might get startup money for protein purification. Mitch asks for permission to use Dr. Jones' name when contacting these new prospects and receives it. Tom Alonzo returns with a copy of the handwritten bidding proposal for Mitch to send to his company as a heads-up for the bid coming from the university's Purchasing Department.

Over the next two weeks, Mitch makes sure that the bid reply from Consolidated is handled promptly. He then checks with Purchasing on the bidding progress. Dr. Jones calls when he receives the bid results, and Mitch makes an appointment the next day to help review them. One HPLC bid can be rejected because it does not meet the specifications. One system for an HPLC system has an additional very expensive detector and can be rejected on the basis of price. Mitch and Dr. Jones then sketch out a justification letter for the Federated 2-D EP and the Consolidated HPLC. Tom takes it to the department secretary for typing on letterhead and then carries it to Purchasing. Mitch goes along, is introduced to the agent who is handling the bid, and asks for permission to pick up the completed university order for the HPLC system so that he can mail it overnight to the company to speed shipping. The purchasing agent agrees, and Mitch gives him his business card so that he can contact him when the order is ready.

When Mitch receives the call from Purchasing, he shows up at the office with a preaddressed envelope, gets his copy of the order, drives to the overnight mail office, and sends the order to Consolidated's order entry system. Next, he books the system in the manufacturing queue and requests shipping expedition, then calls his sales manager to get a release for shipping on the agreed-upon supplies. Finally, he alerts his service representative to the order and the approximate shipping date to schedule installation.

In a quick review of the selling process, Mitch used ADMANO to establish the application with the first laboratory's postdoctoral student. Confirming it with Tom Alonzo, he found the decision maker from the same student but confirms it with Drs. Jones and Alonzo. He determine the amount of money and its availability from Dr. Jones, the next steps in the process from the conversation with Dr. Alonzo; finally, he talked to others involved in the decision. When he first met Drs. Jones and Alonzo, he did a quick HBA on each to determine their motivations for buying. Dr. Jones wanted control of the decision; Dr. Alonso wanted to be able to trust the sales representative and make sure that the system's results could be trusted. The combination of a Driver research director and an Amiable postdoctoral student usually produces a very strong working relationship.

The sale progressed from handling No trust by referral from the laboratory, Mitch's business card, his peptide experience, the company's brochures, and his willingness to help prepare the bidding specification. No need had already been overcome from their research and his visit to the laboratory, and moved to desire by focusing on the research needs and matching them to the advantages described in Consolidated's brochure. No help was overcome as Mitch went over the instrument brochures, pointing out how the advantages of his system fit the department's their purification needs. No hurry was handled by the department's need to get their research up and running and by Mitch's sensitivity to their financial limitations, as shown by his volunteering to donate startup supplies if his system was approved. No money was never an issue since the startup money for the laboratory was sufficient to buy the equipment, was allocated for that purpose, and additional money existed for salaries and operating supplies.

No big close was necessary because the assumptive close arose out of the assistance provided by Mitch in the bidding process. He had become a consultant to the laboratory by defining the detailed specifications locked to his analytical system needed for the bid.

It is important to get into an account as early as possible and to never let a sale dead-end. After the bid was completed, Mitch asked Dr. Jones for referrals to other scientists who might be buying equipment. He also asked for and received permission to use Dr. Jones' name as a referring user. The best time to ask for a referral is when someone had just bought, even if, as in this case, the purchase order has not officially been issued. The next best time is after a system has been installed and is being used successfully. The earlier you get the referral, the sooner it can be used to make another sale.

14

CLOSING THE SALE

Closing the sale is the natural consequence of the sales process. If the sale is not closed, nothing is ordered and the purchase is not made. If the sale is made systematically, the close flows naturally from handling the No hurry problem. In fact, LSCPA., mentioned in Section 13.1.4, is a close because the last two parts are *propose a solution* and *ask for the order*. At this point, the salesperson should be quiet and let the customer talk. The next one who speaks buys! Many salespersons ramble on and end up buying back the instrument that they just sold. This is a lose/lose situation since the salesperson has lost the order and the customer has lost the use of the new equipment.

A myth has been created up in the sales culture about the magic close: If the salesperson uses just the right close, the customer will always buy. This is not so. If the customer trusts the salesperson, needs and wants the equipment, and believes it will help solve problems, he or she will buy it. The only question left is: how soon? The customer is caught on the horns of a dilemma; he or she wants the instrument but does not want to give up the money that was so hard to get. The purpose of the close is to help eliminate the customer's dilemma and get him or her to take action. If only a small amount of money is involved, the problem can be easily overcome. On the other hand, if a considerable investment must be made, a commitment may be much harder. Remember, fear of loss is

Buying and Selling Laboratory Instruments: A Practical Consulting Guide.
By Marvin C. McMaster

more important than the desire for gain. It is easier to put off the decision than to make it immediately. People make only a few big decisions each year, and they are not comfortable about approaching such an unfamiliar subject. The salesperson's job is to create a way for customers to act without hurting themselves.

The first question the salesperson needs to answer is when to make the close. Some salespersons close before they have earned the right to do so, and they then have problems continuing the sales process. This is the dilemma of the salesperson who walks in the door and says, " Good morning. I am from XYZ Company. Your neighbor down the halls says you are looking to buy an HPLC. We sell a really great machine. Why don't you buy it from me right now?" This person has closed but has a very poor chance of getting the order. The customer rightly does not trust him, and he has not earned the right to close or ask for an order. He will have a difficult time getting permission to continue the sale and might even be asked to leave.

A trial close can be made when a customer has accepted a feature or benefit of an instrument and understands how he or she might benefit from it. A close at this point might be rejected, but the rejection simply means that the customer is not yet ready to make a commitment; more information and more convincing are needed.

14.1 ASSUMPTIVE CLOSES

Now that we have shown why closing is necessary, let us look at some successful closes. If you are interested in more information on closing, you might want to read Zig Zigler's book *Secrets of Closing the Sale*.

The *assumptive close* is based on the idea that the salesperson has already done enough to convince the customer of the wisdom of buying now. The salesperson may already have helped the customer write up bidding specification for the instrument. He or she turns to the customer and says, "Can you think of anything else we need to add?" If the customer agrees that everything is included, the salesperson says, "Good. let us walk this over to Purchasing and get the bidding started." If the customer accepts this, the process of buying has started. The sale is closed if the bid is accepted on the salesperson's specifications.

Customers who ask for more information about a feature/benefit that has just been presented are giving a buying signal. The salesperson may ask, "Is this feature really important to you?" If it is, the salesperson can provide more information on that feature and then say, "Why don't we write up an order for this instrument and see how fast I can get it shipped so that you can start using that feature right away. Is that fair?" This is an assumptive close connected to an *urgency close* connected to the sense that a customer always wants to be fair.

Car salespersons commonly use the urgency close in a manipulative form. "Look, this great price is only good until the end of the week, and then we have to go back to the original list price for this fine automobile. You would'nt want to miss this wonderful deal, would you?" This close is based on the idea that avoiding loss is more important than the desire for gain. The way to handle manipulation is to walk away and let the salesperson feel the sense of loss! Not all closes have to be verbal. Body language closes such as walking away can speak louder than words.

The *secondary close* is a form of assumptive close. It gets the customer focused on agreeing to something else, and in the process the customer agrees to buy. The salesperson might say, "Look over this purchase order and see if it looks all right to you; if so, just initial it on the bottom." The customer's focus is on the correctness of the purchase order, but when the customer initials it, he or she has just bought the instrument. Another way of making this close is to say, "If I can get my manager to approve this lower price that you want, are you ready right now to sign this contract?' If the salesperson comes back with approval from some other salesperson, the customer has just bought the salesperson's secondary close.

Putting on the judge's robes close takes advantage of the customer's ego. Before starting the sale, the salesperson says, "Look, I do not know as much about your research and you analytical needs as you do, so I have to ask you to help me. I will show you a number of features about this instrument, and you will have to tell me which ones are important to you. Is that fair?" The salesperson has just transformed the customer into a consultant in the sale. At the end, the salesperson puts the customer back into the judge's robes and asks, "Which features do you feel are most important?" The customer answers and the salesperson says, "Good. Let's get the purchase order cut for this system so that you can get it in here to take advantage of those features right away, okay?" The sale has been closed on the features that the judge has just listed.

When we talked about selling to the Driver personality in discussing HBA, I mentioned an *alternate-choice close*. I told the customer that I would return on a set date with lists of two possible systems, with their benefits and prices, so that he could decide which one he wanted to buy. My assumption was that he will buy one system or the other. Do I care which system he buys? I do not; either one is a completed sale. The customer remains in control while I get the order and the commission, but he gets to keep the instrument when it arrives. Everyone is happy when I get the system up and running with the service man and his technician's help.

14.2 MANIPULATIVE CLOSES

The closes that we have reviewed are designed to help the customer move past indecision and buy. Any of them can be a little manipulative, depending on the intent of the salesperson. The closes I consider truly manipulative are designed to force customers to buy what the salesperson wants them to buy—and buy now—whether it is to the customers' benefit to do that or not. The urgency close and the back-the-hearse-up-to-the door-and-let-them-smell-the-flowers close are more concerned about the welfare of the salesperson than about the welfare of the customer. Their intent is to force the customer to buy now—or else. Closes that invoke fear, anger, greed, or other strong emotions are always manipulative because customers cannot make rational buying decisions when these buttons are pushed. Pretty girls in bikinis placed in a car commercial are not there to promote a rational decision making based on the features and benefits of the specific model of automobile.

14.3 FINAL CLOSING AND THE LOST SALE CLOSE

Closing techniques seem to be the major topic of many sales manuals. These so-called power closes are usually manipulative and are seldom needed if the basic objections are handled properly. Closing is a natural, expected part of the selling process and usually flows from the process of creating urgency to buy now.

One of the myths of selling is that if you use a power close and the customer says "No," there is no way to continue the sale. This seems foolish to any one who has children. They do not quit if they get a "No." They simply wait and then ask again or try the other parent.

Zig Zigler says that if the customer is standing on the table screaming at you, "I would never buy from you. I hate you, I hate your product, and I hate your company!" you can simply say quietly, "Before you make a final decision, let me make sure you know what you are giving up; is that fair?" Generally, the customer says something like "What do you mean? What am I giving up?"

At this point, you have just been given an opportunity to sell again. Think of the five major benefits of your product and say, "Well, you did understand that . . . " (number one). Keep going until the customer has responded to all five. If the customer understood all five benefits, make another appointment at a later time. If the customer reacts positively to any one of the points, ask for the order based on that point.

Does that really work? After hearing about this technique, I was riding with Mark, a salesman who complained about losing a $30,000 order at university we were going to visit. I asked if he wanted the order back. He admitted he did, but said it was impossible because the purchase order had been sent to a competitor five days earlier. I explained the technique, and we practiced it on the two-hour drive from Mark's home to the customer's university office.

We walked in to discuss a separate sale, and Mark said, "Before we start talking about the HPLC system, can I asked you a question?"

The customer said, "Sure."

Mark said, "Before you make a final decision on that scintillation counter you ordered last week, can I make sure you know what you are giving up?"

The customer rocked back in his chair and said, "Giving up? What do you mean, giving up?"

Mark said, "I do a lot of business with the university, and I don't want you to be angry at me because I failed to do my job and let you know that you could get much higher counting efficiency with our system."

The customer said, " I am happy with the other company's efficiency claims, and their system costs less."

Mark said, " You did know that you can mix racks of small and large vials in our system?"

The customer rocked forward. "You mean you can put both small and large vials in the same rack?"

Mark said, "No, but you can mix racks of small and large vials without changing any settings."

The customer said, "That means I could run fifty percent more assays in a single night. If I had know that, I would have bought your counter."

Mark said, "It's not too late. Have they shipped their counter yet? You can always change your mind and cancel a purchase order."

The customer said, "If that's true and you can have a new quote on my desk this afternoon, I will order your system."

The competitor screamed, but the customer bought Mark's $30,000 system—after he had lost the sale. There is no such thing as a lost sale until the salesperson gives up!

Will this always work? Probably not. No technique always works. But at this point, you have little to lose and everything to gain. You are not going to insult a customer by trying to help him, are you?

15

THE LAWS OF SELLING

There are universal principles in selling that all good salespeople use almost instinctively. Looking back over 25 years, I have chosen the ones that have been most useful in my career.

15.1 SALESPEOPLE ARE MADE, NOT BORN

The myth of the born salesperson persists. You can look in an obituary and see the details of a salesperson's life, but show me a record of a newborn salesman. Anyone willing to learn and persist can become a salesperson. However, with the reputation that salespersons have, you might ask, "But who would select such a career?"

We are a nation of salespeople. Most people came to America because of the opportunity for religious freedom and to take advantage of the bounty of this new land. We are the wealthiest nation in the world because we can make and sell anything from water in bottles to automobiles and airplanes that span the nation and the world. We have sent men to the moon, and a salesman sold the idea to Congress and the American people.

The day I walked out of the research laboratory with a Ph.D. in organic chemistry and 12 years of experience in biochemistry, analytical chemistry, and process development my income was 20% higher than it was

Buying and Selling Laboratory Instruments: A Practical Consulting Guide.
By Marvin C. McMaster

as a research/process chemist. After five years of sales experience, I was making 50% more than the average chemist. I made more use of my technical training in helping to solve laboratory problems as a sales consultant than I ever did in a research laboratory as a bench or process chemist.

Learn and persist. There are libraries of books that teach you to sell and help you develop a relational personality. Most of these books are a tenth of the cost of technical textbooks. Training in almost any other field will translate into selling expertise and make you a unique and marketable commodity. Unemployment in sales is almost unheard of unless the person is seeking a better employment opportunity. But if you want to be a profitable salesperson, become a professional salesperson. Study your field and practice your craft and profession. Just get better!

15.2 YOU ONLY HAVE ONE CHANCE TO MAKE A GOOD FIRST IMPRESSION

The first call on a customer is critical. Know as much about the customer and his or her job as you can before you walk in the door. Make eye contact and learn how to smile when you decide to smile, not as a response to someone else's smile. Use ADMANO and HBA to find out the customer's interests and focus your conversation accordingly. Ask open questions to find out what the customer is interested in and what he or she wants. Pace the customer's emotional style. Establish yourself and your company as trustworthy. Study and learn to use body language as a nonverbal form of conversation to assure customers that you are there to serve them and help them get the products they need to do their job well. Dress well and be personally clean. Use the golden rule: Be the person you would want to have call on and serve you. Be enthusiastic! Tell yourself that you are even if you are not. The customer is not interested in your problems; leave them at home. The customer has problems that you are there to help solve.

The first 30 seconds and the first five minutes in a sales call are critical. Get the conversation on point. With an Amiable, keeping the conversation on kid, guns, and dogs may not seem like part of the selling process, but it is. If the customer offers coffee, take it. You don't have to drink it, and it gives you something to do with your hands. I learned to drink coffee black so that I did not have to take critical time away from the customer and his interests adding sugar and creamer.

Have a plan for the call and try to adhere to it, but adjust to the direction in which the customer wants to go. Remember, you are a consultant, there to help solve the customer's problems. Control your ego and your speech. Bring worthwhile ideas and stories. Leave the gossip, profanity, and raw jokes to the sales con artist.

15.3 SALESPERSONS ASK QUESTIONS, NOT MAKE STATEMENTS

A statement invites argument and has to be defended. Questions are asked to seek information and draw customers out. You do not have to answer a question. Ask another question to continue to acquire information. If a question comes up three times, it is important to the customer. Note it and answer it, but immediately ask another open question to continue the inquiry process. An answer is like a statement and does not continue the process of acquiring information. "Oh?" is a question that asks for more information. It asks the person to explain and defend his or her last statement. "Why do you feel that way?" is a similar type of response. A salesperson can learn to become a question-generating machine by practicing on his or her spouse, relatives, and friends. People never really object to questions since the questions are about their favorite subject.

Open questions like "Oh?" and "Are you interested in expanding your laboratory?" are designed to get persons talking about themselves and their interests. They are not designed to elicit specific responses. Open questions are used to fish for information. They begin with the words *who*, *what*, *where*, *when*, and *how*.

Closed questions are specific, guide people to a response, and can usually be answered with "Yes," "No," or a number—for instance, "How much money do you have budgeted for this project?" or "When do you expect your funding will be released?" Here you are looking for a specific response from the customer, and you will have to follow up with an open or closed question to continue to ask the customer to volunteer additional information.

15.4 FEAR OF LOSS IS MORE IMPORTANT THAN DESIRE FOR GAIN

People do not want to lose or give up something. They will pass up a chance to gain something new before they will sacrifice something they already have even if the thing to be gained is much more attractive and valuable. This seems to be an instinctive reaction rather than a logical one.

It is not a matter of weighing the two items. As Nikita Krushshev said at the United Nations, "What is mine is mine; what is yours is negotiable." We want to hang on to what is ours. This is the basis of the monkey trap. Trappers will drill a hole in the bottom of a milk bottle and fasten the bottle to the ground with a knotted rope pulled through the hole. They will put bait in the bottle and leave it out for a monkey. The monkey will be able to squeeze a hand into the bottle and grasp the bait but will not be able to pull the fist holding the bait back out of the bottle. His desire for what he has will not let him get the freedom he wants.

This is the basis of the lost sale close described in the previous chapter. The customer had placed a purchase order. For him the order was completed, and people do not like to change decisions. It is too much trouble to make decisions in the first place. *Never ask customers to make a new decision without giving them new information!* They were willing to listen because they wanted to make sure that they were not giving something up. This gives the salesperson a chance to offer new information or information the prospect failed to hear before. The sale is made on the basis of this new information. You are telling customers that they are stupid or that they made a bad decision if you ask them to make a new decision without giving them new information or the same information in a new form.

15.5 IF YOU DO NOT ASK, THE ANSWER IS AUTOMATICALLY NO

Zig Ziglers's younger brother, Judge Zigler, produced a sales tape titled *Timid Salesmen Have Skinny Kids*. A salesperson who will not ask for the order is an oxymoron. If the salesperson will not close the sale, there is no sale, and the salesperson keeps the instrument and has just wasted the customer's time and done him or her a disservice. A salesperson who will not close is a professional visitor.

No good sales manager will ever hire a salesperson who does not ask for the job. The salesperson will not close a sale if he or she will not close a manager about a sales position. I was not aware of this when I interviewed for my first sales job. I had acquired a castoff HPLC system in my plant troubleshooting job and had used it to solve two major problems with very little training. I was so impressed that I cornered the HPLC salesman on his next visit, got telephone numbers, and acquired his sales manager's name and number. I called him and set up an appointment for an interview. He flew me to the company headquarters, gave me the tour, had me talk to a number of people, and then invited me to his office. He said, "Well Marvin, what did you think?"

I answered, "I don't know about you, but I have already made up my mind. I'm just waiting for you to do something." I did not know it, but that was a close. The company offered me a job, and I took it. They taught me to sell their hardware, and I have been selling in St. Louis for 35 years. When I worked in a research laboratory at DuPont, someone said that I might want to consider a sales position but warned me that it was a one-way career path. At the time this discouraged me, but I found later that he was right. I have found this career so satisfying and profitable that I would have been crazy to give it up.

This subject is also very important to potential customers. Customers need to ask for the best equipment for their research or the answer is "No." Customers need to ask for the best discount they can get or the answer is "No." Customers need to ask for the best service they can get or the answer is "No." Life is too important to slide through it diffidently. Timid customers have small research laboratories. That is the second part of the win/win relationship; insist on what you deserve.

15.6 LISTEN MORE THAN YOU TALK

It's important to listen at least twice as much as you talk. Ask question and then listen to the answers. When I got into sales training, I was so enthusiastic about the material that I kept interrupting and asking questions. The trainer finally made me take a listening exam. I am sure the company was trying to tell me to be quiet, but I was confused. Later, I learned that I got the highest score in the history of the company on that test. I did not reveal that I had just read *The Memory Book*, by Harry Lorayne and Jerry Lucus, and used the listening techniques that I had learned there.

Selling is about asking questions, listening to the answers, and then applying them to help customers get what they need to be successful. People are starving for someone to listen to them. In his book *How to Win Friends and Influence People*, Dale Carnegie says that he was at a party and was introduced to a botanist. Carnegie said, "Really? I have always wanted to know how to be a gardener." For the next two hours, the man told him everything he knew about cultivating plants, and when he left, he complimented the hostess for inviting such an intelligent and interesting conversationalist as Mr. Carnegie. Dale Carnegie said, "I had asked only one question the whole night, but I listened attentively."

After I had entered sales, when my wife and I shopped for furniture, I shopped for salespeople. I asked them how they came to be in sales, and I listened to the answer. They told me their life stories. If they began to slow down, I asked another question about something they had said and they were off again. On a sales call, ask the customer for permission to take notes. This is a tremendous compliment; it implies that what the customer is saying is important. It is also important to you if you are planning to make the sale. Remember, on a selling call, when you talk, you lose ground; when the customer talks, you both win.

15.7 OBJECTIONS ARE A SIGN OF INTEREST

People do not object when they are not interested. They do not buy when they are not interested. New salespeople often have problems answering objections. They interpret the objection as a personal rejection rather than as a question about a feature or a claim about the instrument they are selling. They have been told or have heard that *the customer is always right*. Correctly, they do not want to argue with a customer. As a result, they often feel intimidated, unsure how to proceed, and do not pursue the sale. Customers are always right, but they are often confused at the moment. They only have to be right when they say "Yes" at the end of the sale.

Instead, when a customer expresses a strong objection, the salesperson should think, "This person is interested! He (or she) feels strongly enough about this feature or idea to argue with me, so it is probably the one thing about my equipment in which he (or she) is most interested. I can probably close the sale on it if I answer this objection and show how important it is to the client's success."

Answering all the minor objections can slow the sales process. Note them and tell the customer that will go over them when you finish explaining the features. If the same objection comes up a few more times, stop, answer it, and try a trial close on it because it is a major question. If the client brushes off the close, go back to your presentation, but note that objection and use it at the end of the sale.

15.8 DO NOT ARGUE, ASK FOR CLARIFICATION

Customers often object strongly to things they do not understand or disbelieve. It's natural to defend things that you know to be true. Don't! This is the point where "feel, felt, found" becomes very valuable. Ask customers to help you understand what they believe: "I'm not sure I understand what you mean. Could you clarify it for me? I try to make the best presentation I can, and if I have made a mistake, I certainly want to correct it before I tell another customer something wrong. Is that fair? You would be willing to help me, wouldn't you?"

You are putting the judge's robes on the customer. Customers will go back over what they said, rephrase it, and tell you again. In thinking back over the idea, they may discover that they made a mistake. If not, you can use "feel, felt, found" to explain the problem and the solution. Customers may surprise you and show you that you were indeed wrong. You will probably find that you have made a friend by letting them help you if you thank them for the correction.

By the way, there is magic in the phrase "help you." Use it often in your presentation if you are truly in the customer's office or laboratory to provide help. People understand that you know more about your equipment than they do. If you sincerely offer to help them, you become a professional rather than the con artist they expected a salesperson to be. Their understanding is that consultants help, whereas salespeople take away.

15.9 BODY LANGUAGE CAN DEFUSE SALES TENSION

Study body language in motion as an aid to understanding and as a tool for removing emotional tension from a sales call. Body language makes up about 70% of human communication. Words, intonations, and facial expressions make up the remainder. Most people use and read body language subconsciously; learn to do it consciously. People believe you when they see congruence between your words, face, and body language. You must learn to consciously interpret the body language of other people and your own. When I first studied body language, I was taught to read static body language. As I used this skill in sales, I found that body language in motion was much more reliable. You will need to probe further with questions when customers' body language disagrees with what they are saying.

I had startling proof of the power of body language in defusing sales tension. My district manager, Lou, was traveling with me on sales calls. We went into a scheduled visit with current customers in my most important account. I could tell immediately that there was a problem. Both customers sat on opposite sides of the table from my manager and me. Their arms were crossed, they sat back in their chairs, they frowned, and they described their problems in no uncertain terms. Thoughts of disastrous lost future sales flashed through my mind.

My manager, Lou, leaned forward with his hands on the table and listened carefully. When the customers finished describing their problems, he restated them quietly, in a low voice, and then asked a couple of questions in the same voice. The customers leaned forward to hear what he was saying, their hands came down on the table, and their expressions turned friendlier. Finally, the lead speaker for the customers said, "Those are interesting questions. We need to think about them for a second. Let's take a break and get some sodas."

They escorted us to the soda machine, bought drinks for us, and we returned to the table. When we sat down, the four of us were scattered around the table and the problems were quickly resolved. After we finished and left, I asked Lou if he knew what he did to defuse the situation. "All I did was to summarize their concerns and offer a couple of suggestions," he said.

"Are you aware that when you did that, that you dropped your voice, leaned forward toward them, and placed your hands on the table, and that they matched you and the whole situation changed?" I asked him.

"Really?" he said. "I did not notice."

Lou had sold in the field for years and had only recently been promoted. His body language had been acquired on the job and was adjusted to fit the sales situation. By dropping his voice and leaning forward, he had drawn the customers forward. They dropped their aggressive crossed-arms position as they leaned forward to listen, and they put their hands on the table. When we paused for the soda break, they returned to the table and broke the confrontational arrangement of *us on one side versus you on the other*. After these changes, everything proceeded smoothly. Lou had done these things through experience and generated body language changes. I observed this procedure and tried it in other sales situation. It worked the same way to ease the tension generated by first visits, emerging problems, and customer disagreements and objections.

I have said that body language in motion is more significant than static body language. Crossing the arms and sitting back in a chair can indicate hostility, but it also can mean that the person is cold. Women often cross their arms to protect themselves when they feel cold or threatened. However, when you ask a question and the customer responds by leaning back in the chair and crossing his or her arms, watch out! The customer has just signaled that disagreement with what you have just said, especially if he or she frowns and looks to the side. I saw this in a courtroom when a witness gave a similar signal, and what he said later proved to be a lie. Watch for changes in people's body language that are congruent with their words and expressions to help you know what is going on. Get a book and read about body language; it will improve your sales.

15.10 EMOTIONAL BUYING AND LOGICAL JUSTIFICATION

In Chapter 14, I said that people make emotional decisions to buy and justify them logically. Our ears allow emotional access to the brain. Salespeople are educated through their ears and are taught to gain major emotional access to customers through their ears. Open and closed questions access the customer's ears while allowing the salesperson to maintain the momentum of the conversation. Other senses allow emotional access in a sale; the smell of fresh-cooked food makes it attractive, the sound and smell of fresh popcorn draws us in, the touch and feel and smell of fine leather sells cars. Even though the eyes give us access to the logical, statistical side of the brain, they also allow access to the emotional side of the brain via bright colors and rapid movement. After the decision to buy is completed, we call up the pictures, specification, tables, and calculations we were shown to justify the decision we have already made and to prove that it is the only logical decision we could have made. The scientific world sense we feel around us makes us need this logical confirmation. Even words support the emotional decision. Logically, we *think* we need to buy. Emotionally, we *feel* we want to buy. A good salesperson does not ask you if you *think* this is the right decision. Instead, he or she asks you how you *feel* about your new car and helps you picture gliding down the Pacific Coast Highway in the comfort of a powerful, shiny new car. All this emotional appeal eases you into the decision.

15.11 PEOPLE WANT TO BE FAIR

Once the trust barrier is lowered in a sale, people want to be fair. They instinctively want an equal win/win relationship. They understand that you need a profit to make a living and stay in business. But they also do not want to seem or feel foolish because they paid too much or got too little for their money. A good salesperson will always compliment a previous buying decision that the customer has made. Customers want to make sure that they get a good deal this time if you tell them that they should have gotten a better deal last time. When a salesperson proposes a particular system configuration, he or she turns the statement into a question by adding, "Is that fair?" because the salesperson knows that the customer wants to cooperate in completing the sale.

15.12 HONESTY IS GOOD BUSINESS

You probably have heard that honesty is good policy. In business, honesty is not a policy but a way of life. Honesty is like virginity; once lost, always lost. A half-truth is a whole lie. Lies, like chickens, always come home to roost. Once the salesperson loses a customer's trust, it is almost impossible to regain, especially if the customer is an Amiable. This is not about morality; it is about business and profitability, because eventually it will affect your bottom line.

15.13 NEVER CRITICIZE AN OPPONENT

A common problem with new salespeople is that they criticize the opposition, thinking to build themselves up by tearing down competitors. You always lose stature in your customer's eyes by attacking your competitors. This tactic makes you look small when you belittle someone else. Build your own integrity by giving your customers the very best deal you can find for them and let your competitors try to live up to the standard you are creating. Be a pro, not a con.

15.14 TANSTAAFL

The American delusion is the belief that you can get something for nothing. It is an illusion that fuels all lotteries, Ponzi schemes, raffles, give-aways, scratch-off cards, and similar con games. In fact, you get what you pay for if you are lucky. Something for nothing is a perpetual-motion machine and violates natural law. More often, you get less than you pay for because someone has cheated you. Hoping to break even is the best you can expect. The summary statement TANSTAAFL, coined by Robert Heinlein, stands for "There ain't no such thing as a free lunch." The comedies of life are full of people learning this painful lesson. A win/win sales relationship has no place for a free-lunch delusion.

15.15 EXPLAINING QUALITY OR APOLOGIZING FOR THE PRICE

Offer your customer the *best fit* for his or her needs. Sell big because it is easier to sell down rather than selling up. Offer your customer the less expensive 55-gallon option instead of a 5-gallon pail if he or she will be using the product repeatedly. If you start with the 5-gallons size, it is very hard to work back up to the 55-gallon product. Justify the larger size by pointing out that it is cheaper to buy this way. You might offer to absorb the shipping cost on the larger instrument if the customer still needs to be convinced.

You will often find that the less expensive system you sold is inadequate for your customer's future needs if you try to fight a pricing battle. It is easier in the long run to justify a more expensive, more capable system than it is to apologize repeatedly for the lower price of an inferior product.

15.16 THE WORD *SALES* COMES FROM SERVING

The quality of salespeople would be better if they all were taught, from the sales manager down, that selling means serving. The word *selling*, according to Zig Zigler, originates in the Norwegian word *selye*, meaning "to serve." Selling is actually very simple. It involves helping people find what they want, showing them how to get it, and letting them buy from you. The next time you want to buy something, you probably will go back to someone who served you and helped you buy at a good price in the past. That is called building an account with repeat business, and it is what real selling is all about.

16

HANDLING PROBLEMS

A problem often occurs during the sale or after a system is installed. How the salesperson handles it determines his or her professional quality. Attitude is critical in producing a positive outcome. Any challenge can be either a problem or an opportunity. Handling it requires resources either in your home base or in the manufacturer's headquarters.

As a salesperson in training, you should spend time in-house locating a resource person who can help you expedite orders and shipping. This might be a member of management team, but most commonly it is a secretary or a customer support person. Showing appreciation for this help is critical; these people should be cultivated, complimented, and rewarded when they dig you out of a hole.

Field service personnel are your best friends in the company. When things go wrong, they are your first line of defense, but you should have at least one competent source of service advice in your manufacturer's home location. You may have trouble getting your field service person on the phone when he or she is on a call, and the in-house technical person is usually more readily available. This person does not have to be in the service department; often, I have found that research and manufacturing personnel are better at solving unusual problems.

With luck, you will also find an in-house person who is extremely well organized and competent. If you need an answer ranging from catalog

Buying and Selling Laboratory Instruments: A Practical Consulting Guide.
By Marvin C. McMaster

information to application details, you can trust this person to get back to you with the information. I have found a few of these people in my sales career, and they are invaluable. They are walking Day-Timers, and they know how to prioritize everything. One such person I knew in my first position had started as a teenager working summers for the company, had become a company application chemist after college, and was moved into the customer support office. He knew more about the details of the company, the instruments, and the work they did than the company president. Eventually, the company recognized his competence appointed him as a vice-president long before this position became a promotional burial ground in many companies. Even after I had left the company, I was able to contact him for critical information.

16.1 WARRANTIES AND CUSTOMER EXPECTATIONS

Customers have the right to expect that the instrument will work and will be capable of handling their application problems. A salesperson's job is to help meet this expectation. A major problem occurs when a bad instrument slips through quality control and become a threat to future sales in the customer's institution. Your field service technician should usually handle hardware problems, but ultimately, account responsibility rests with the salesperson. If the whole system is a lemon, the company should be willing to replace it for a period of time specified in the warranty. There is little you can do to influence this situation except to make sure that you are working for companies that are pros, not cons.

The salesperson is the face of the company that the customer sees, and you should fully understand the customer's needs. This can be difficult if you are working in a widespread territory that brings you in close contact with all of your customers only a few times a year. Make sure that all of your customers have rapid access to you. Usually this is through the company telephone exchange and message service. It is important to return calls promptly so that problems are not allowed to fester.

As I discussed in Chapter 7 for a period of time, my local service representative was promoted to an in-house management position. Suddenly my support for installed instruments was being sent in from neighboring territories. About 80% of the time, I could handle any problems.

I sent my boss and the area service manager a report stating that I had take care of the service problems. They quickly replaced my local service representative. The goodwill that my actions had generated continued even after the new service technician was in the field. I got calls from the new service man telling me that I needed to see specific persons getting ready to buy a new system from me.

16.2 DEALING WITH A LEMON

No manufacturer is perfect. Bad instruments occasionally come off a production line, but they are usually picked up in house if good quality control procedures are in place. Modular systems have the advantage that if a bad component does slip through quality control, the bad component alone can be replaced instead of the whole system. Again, the sales representative has ultimate account responsibility and should get involved early, help diagnose the problem, and initiate the replacement. Such problems usually happen soon after installation and are easily covered under the manufacturer's warranty. Commonly, field service personnel will get involved and will struggle to make repairs over a period of time. The problem will extend past the warranty period, and an argument will break out as to whether the repair is covered or not. This problem can be avoided if the salesperson quickly gets involved, insists on the replacement, and avoids a negative sales reference.

Not all problems are instrument based. It is important to realize that service people are trained to fix hardware and electronic problems. Chemistry problems are beyond their area of expertise and can greatly delay a service solution. Early in my sales career, a customer called and threatened to send back a $30,000 HPLC system. The chromatograph produced from the system was nonreproducible, and the customer was irate. I went to the customer's laboratory to look at the problem. Immediately, I realized that the situation was a water problem, not a chromatography problem. Naively, I told the customer that his water was bad, which was a serious mistake. He informed me that his triple-distilled water was good enough for his biochemical incubations, and it was good enough for his chromatography. We slowly replaced every component in his system without solving the problem. Just before it seemed that we would have to take back the system and lose the customer, he went to a local university and borrowed HPLC quality water from one of its laboratories. He washed out his column, switched to the HPLC water, and the problem disappeared forever. I actually received a telephone call apologizing for this problem. Many organic compounds can codistilled with the water and were accumulating on the head of the column, changing the chromatography in less than a day. A very valuable lesson was learned at a high cost in time to everyone involved

16.3 INSTRUMENT SUCCESS GOALS

A successful sale does not happen just because you have won the purchase order. A successful sale is made up of rapid shipping, installation, instrument turn-on, and a successful application run by the customer or his or her technician. Regular instrument use is the goal of every sale.

The first step in achieving this goal is to get the purchase order to the company and the order booked into the manufacturing queue. The salesperson gets credit from the sale toward the quota when the order is booked. Purchasing departments never seem to have a sense of urgency about submitting purchase orders. Some departments will let you submit the purchase order for them to save postage, and I have used priority mail to get the order booked quickly. An in-house contact person at your company can speed up booking, shipping, and making your sales manager happy. He or she knows only too well that anything that speeds up the order can help prevent lost sales from disasters that occasionally happen, such as cancelled grant money as budgetary constraints occur. A bird in the hand is worth two in the Purchasing Department.

Make sure that the customer checks for obvious damage and missing parts once everything has arrived in the customer's laboratory in large foam-lined cube boxes. It is important to look for damaged corners and forklift marks on the sides of boxes. Have the customer call the company's insurer if any boxes show obvious external damage. The warranty process becomes much smoother if this simple call is made immediately.

Notify your service technician as soon as the instrument is shipped so that he can book the installation with the customer. Once the system is up and running, I try to schedule a follow-up visit to make sure that the customer has a staff member using the machine. In some laboratories, this is no problem; if the instrument is installed, someone is using it. It is with the first-time users, who just let it sit there while they admire it, that the instrument can become a negative referral. I try to visit these accounts as soon as possible with my instrument startup protocol in one hand and a reservation for a training school for the operator in the other hand.

I began teaching at instrument schools in my territory when I started working for a manufacturer that did not offer such training. I persuaded my boss to give me time to teach the courses in our office using my demonstration hardware. I taught the basics and the theory of operation the first day and performed hands-on demonstrations the second day so that every customer had the opportunity to make one injection in an HPLC and get a chromatogram. Later, we gave two tuition wavers for each instrument sale as an add-on benefit. The schools were so popular that eventually we took them on the road. In three years, I did 33 two-day courses in St. Louis and in sales accounts in seven surrounding states. These free schools crushed the competitors that had no such offerings and eventually allowed me to do consulting courses in continuing education departments in local universities and institutions. But the basic goal of all of this effort was to get instruments up and running to act as referrals for new sales.

16.4 PROVIDING APPLICATION SUPPORT

Some manufacturers maintain application laboratories that help customers by developing protocols for the customers' separations to help them use the instruments the company sold them. These laboratories are not designed to do customers' research for them. They are designed to give customers a starting point for using the instrument. Anything that can get the system up and running is a benefit to the salesperson. Another function of the application laboratory is to provide method development research to speed the customer's application or work around a difficult separation problem that the customer has not been able to solve. Unique separations codeveloped by the customer and the application laboratory can lead to joint publications that indirectly benefit the whole user base.

16.5 TERRITORY MANAGEMENT

Poor management of a territory is the main problem leading to poor sales production. New salespeople run their territory reactively instead of proactively. They chase after new leads, competitive activity, and problems that arise instead of working systematically around their assigned area, staying in communication with their installed customer base, and doing only qualified demonstrations. The new representative ends up with an inflated monthly bill for gasoline, mileage on their car, airfare, and wear and tear on their body and nerves.

It is important to analyze where the sales in your territory came from in the past. You will find that 20% of your accounts have bought 80% of the hardware in your territory. Use this as a guide in deciding how to spend your time. Understand that old accounts may stop buying and new accounts in new disciplines may spring up. It is important to manage a territory systematically while keeping an ear open for new business. When new leads open up from technical magazines and trade show advertisements, follow up on them rapidly by telephone and e-mail, and then work them into your territorial planning with scheduled appointments. The only exceptions are people with money in hand and a need to spend it immediately. These people often can be handled with a telephone call, referrals, a faxed price quotation, and a guarantee of help with installation and application development on your regularly scheduled visit to their institution.

16.6 CONFIDENTIALITY

During my career in sales, I have worked with many customers in refining or developing new separations. I have been in and out of their laboratories as I sought new sales and we discussed ideas.

Confidentiality is very important in the race to publish and to maintain industrial secrets. Many large firms require the salesperson to sign a confidentiality agreement. Where possible I carried ideas from laboratory to laboratory and combining them into systems. I was careful to stress that I did so only when I could do it without violating my customer's confidentiality. I explained that where possible, I combined the systems into the information I used to create the slides for my courses and books to help all of my customers.

16.7 SALES INTEGRITY

Integrity in a sales career is critical for long-term success. It is made up of many of the things we have already discussed: honesty, consistency, service, and a consular win/win attitude. But it goes much deeper. Salespersons must truly care for the welfare of their customers and the instruments they sell. A salesperson can sell anything, but a professional salesperson wants to work for a company that also cares about their customers and the equipment they manufacture. You are continuously apologizing if you are selling poor equipment. You may have to explain the quality and value of good equipment to justify a higher price, but this is much more constructive to building personal integrity, sales expertise, and a fully satisfying career.

The single most important sentence in your sales vocabulary is "I will help you!" This is exactly why people come to you as a salesperson. They need your help and expertise. They have a problem that must be solved, and they cannot do it themselves.

I would have included this sentence among the laws in Chapter 15 as one of the keys to effective selling. Unfortunately, nothing will raise the barrier to trust faster than this sentence because it has been so widely misused. There is no single yardstick of a person's integrity better than these four words. Do not use them unless you mean exactly what you say. Trying will not do it. You must help customers or not. This sentence is the heart and soul of selling as service. Fulfilling the promise is the true measure of your character as a professional salesperson.

APPENDIX A

FREQUENTLY ASKED QUESTIONS

A.1 FREQUENTLY ASKED PURCHASING QUESTIONS

1. **How can I be sure that I am getting the best price?**
   There is never a guarantee, but if you are paying the list price from the manufacturer's catalog or price quotation, it is not the lowest price. The GSA receives the best price and insists on it. Find a friend or colleague working in a government laboratory and ask to see his *GSA price list*. Putting your instrument out for bid will help you get the best price, but be sure that you are comparing apples with apples. Some companies' instruments will include training and application support that may not show up on a bid system form. Try to include these in the bidding specifications if they are important to you. Demonstration systems may be offered at a demo discount, but if they are truly demonstration systems, they may have some serious problems. The best thing to do is to negotiate a discount in the form of *supplies, consumables, training*, and *support*.
2. **How do I know exactly what I need to buy?**
   Look at *technical articles* featuring the same type of equipment and see what that facility used. Get brochures from major manufacturers and compare the specifications they offer to the needs of your

Buying and Selling Laboratory Instruments: A Practical Consulting Guide.
By Marvin C. McMaster

laboratory projects. Ask a *colleague* who is using this type of equipment what he or she recommends. Find a *good sales representative* whom you can trust and tell this person what you are doing. Ask for a recommendation for equipment. If it sounds too expensive, ask if you really need all that equipment to run your research projects.

3. **What happens if my research goes in another direction?**
4. **How can I expand or change this system to match the state of the art?**
   Try to build some flexibility into the system. Modular systems are more expensive than systems with everything in a box, but they are easier to repair and expand with state-of-the-art accessories. Sometimes it is easier to dedicate the old system to an often-used application and write a new grant application for a more modern, more flexible system after you gain experience and are more knowledgeable about the technique.
5. **How do I develop grant and bidding specifications?**
   Again, technical articles and manufacturers' brochures can be used to develop your proposals and define exactly the equipment you will need. Be careful not to lock out innovative new accessories, especially ones that increase detection sensitivity or decrease analysis time.
6. **What help can I expect from the companies? From the salespersons?**
   Most good companies that have been around for a while are interested in building long-term relationships with their customers. Ask for their operating manuals and review them. Find out if the company offers an operator training school. Ask for a list of people who are using the system. Find out if anyone in your company is using the system. If the price of the company's system is much less than the prices other manufacturers, find out why. There is no such thing as a free lunch, especially in competitive sales situations.
7. **What problems do I need to watch out for?**
   Problems include prices that are too high or too low, not enough capability or expandability, a system that is too slow or not sensitive enough to achieve your purpose, and a system that has a reputation for unreliability in the technical community.
8. **How quickly can I get the system up and running?**
   With the price you pay for laboratory instrumentation, it is worth while to take the time to get it right. If you have to write a grant proposal and then go out on bid, it might take six months to a year to get a system. Once the money is in hand, it still might take one to

two months if you have to wait for bids to be answered and response studies to make sure that the system fits your needs. On the other hand, if your boss hands you a lot of money and tells you to spend it, and if you have a sales representative who can walk the order through his Purchasing, Shipping, and Installation departments you could get the system up and running in less than two weeks.

A.2 FREQUENTLY ASKED QUESTIONS ABOUT NEW INSTRUMENTS

1. **What do I do when my instrument arrives?**
   Check all the foam boxes for *shipping damage* and call the manufacturer to let the insurer know if anything is bent or if there are holes in the boxes. Call the manufacturer to get an *installation date* locked in. If the instrument is user installable, open the box, carefully turn it upside down, and slide the instrument out on the ground. Set it up, *plug it in*, and see if any lights come on. Get a *surge protection power strip* and plug all components into a common power source. Run a standard that is stable and record the initial values for sensitivity and range so that you will have *established a baseline* to return to when problems occur. Make an initial run as soon as possible
2. **How do I get my staff ready to run the instrument?**
   Find training for your staff by talking to the sales representative who sold you the system. Local universities, manufacturers, and many technical organizations offer training seminars and workshops.
3. **What supplies and tools do I need?**
   Again, go to the literature and see what is recommended in articles by persons who use this type of equipment. If a manual was supplied with the system, review it for recommendations. Look at the manufacturer's web site to see if it has a list of recommendations. Talk to the service technician who does the installation and see if he has a list of recommended tools and spare parts. A good installer will have such a list because it makes his job easier.
4. **Where are the service and sales personnel?**
   That is an interesting question, but it is a little late to be asking it after the equipment arrives. Some companies feel little obligation to the customer after the system is delivered and paid for. If you try to return it, they will want to charge you a 15% restocking fee. It is best to deal with this issue before purchasing the instrument. Talk to other people who have purchased from this company. Make sure that the

salesperson will call again before your next purchase opportunity. Account servicing is part of the responsibility of a professional sales and service company. If the company fails to do this, contact your state attorney general and the local Better Business Bureau.

5. **How soon will I have to replace this system?**
   Laboratory instruments have no specific lifetime. I know of 20-year old equipment that is still working well. Staying current with the state of the art in instrumentation can be harrowing since advances in the field occur very rapidly. The secret is to upgrade software and replace detectors with more sensitive modules until your instrument becomes obsolete. Then dedicate that system to a specific application and write a grant for a new system.
6. **How do I keep the system up and running? Where is the manual?**
   Manuals shipped with the system seem to have disappeared in the last few years. Manufacturers apparently do not believe that customers read them. They have started putting manuals in software help screens and in on-site web sites. If there is any documentation with the system, check it for a web site address or a company telephone number. A good sales representative will be able to help you find a manual and perhaps even supply you with a hard-copy manual for you laboratory. Let the manufacturer know that a laboratory copy of the instrument manual is a condition of the sale.
7. **Why is there smoke coming out of the back of the system?**
   I have heard this question so often that I decided to include it here. Smoke is a sign that something has gone dramatically wrong. Surge protectors are important and will help. Lightning strikes can fry electronic components, but even the way some companies expand their building addition's electrical systems can cause problems. I have seen a demonstration HPLC pump burn up because of a power surge in a company's electrical system.

A.3 FREQUENTLY ASKED QUESTIONS ABOUT THE SELLING PROCESS

1. **Who is this sales person? This company? How did this person get my name?**
   Most commonly, the first contact with a prospective customer occurs when this person answers a check-off card in a technical magazines or requests a brochure from your company. When you show up, he or she has no idea who you are or why you have come. All this person

wanted was the literature. The salesperson's first job is to introduce himself or herself, the person's company, and the instrument(s) he or she is selling. Next, the salesperson needs to find out what the prospect is doing, whether he or she needs any equipment, and if the prospect has any money or any chance of getting money.

2. **Why does the salesperson ask so many questions?**
   The salesperson is trying to find out if there is any way he or she can help you or if he or she is just wasting your time. Questions are the salesperson's method of choice for gathering information and positioning the equipment for sale. A salesperson at this point is a walking brochure with a search function. Ask this person questions in turn and you can get the information you need faster than from any brochure or company web site. You might even find something that will help you solve your research problems.
3. **Why should I tell the salesperson how much money I have to spend?**
   If you do not have money or prospects of getting money, there is nothing the salesperson can do to help you. Of course, you probably will not be doing much research under these financial constraints. If the salesperson knows what budget you have and the applications you are trying to run, he or she can show you the available equipment that might be able to help solve your problems.
4. **How is this machine going to solve my research problems?**
   That is exactly why the salesperson asked you about your applications. His or her job is to understand how the equipment for sale works and how it can be used in your laboratory to achieve those applications. The salesperson cannot solve your problems, but there is a good possibility that the instrument for sale may be able to help you solve them. His or her job is to show you what it can do; your job is to see how it fits your research needs.
5. **Why should I buy it now instead of that gene analyzer?**
   Maybe you should not buy the salesperson's equipment if you really need something else. That is what he or she is in your office or laboratory to help establish. That is the purpose of asking all the questions. If there is no need and the salesperson cannot help you, he or she should thank you and leave. Some salespersons have a quota to make and only limited equipment to sell, and they will overstay their welcome. It has been said that if the only tool you have is a hammer, every problem looks like a nail.
6. **That is a lot of money. Can't I get this cheaper from someone else?**

I am sure you can. Someone is always ready to sell you cheap equipment. You can buy ill-fitting equipment, old demonstration equipment, and junk that fails to fit your needs. If you are lucky, you will get what you pay for. If you are unlucky, you will buy something totally inappropriate for your needs. The salesperson's job is to sell you what you need to succeed in solving your laboratory problems.

7. **Who will repair this instrument when it breaks?**
   A company with a professional service department and excellent field service personnel are the reasons some companies are so successful. They make excellent-quality equipment with a very low failure rate, but when problems occur, they have people in the field who can fix the equipment quickly. With many companies, when things go wrong with your equipment, it is up to you to get it back to work. Suddenly, your system turns from a research solution to a research problem.
8. **Doesn't my colleague have one of these systems that I could borrow?**
   Yes, he does. He has two systems that his research students are using full-time to discover answers to his research problems and to generate all those papers he is publishing. If they work that well for him, shouldn't you be getting your own system as quickly as possible to start turning out answers for your own students and research papers?

APPENDIX B

MEMORY AIDS, FIGURES, AND TABLES

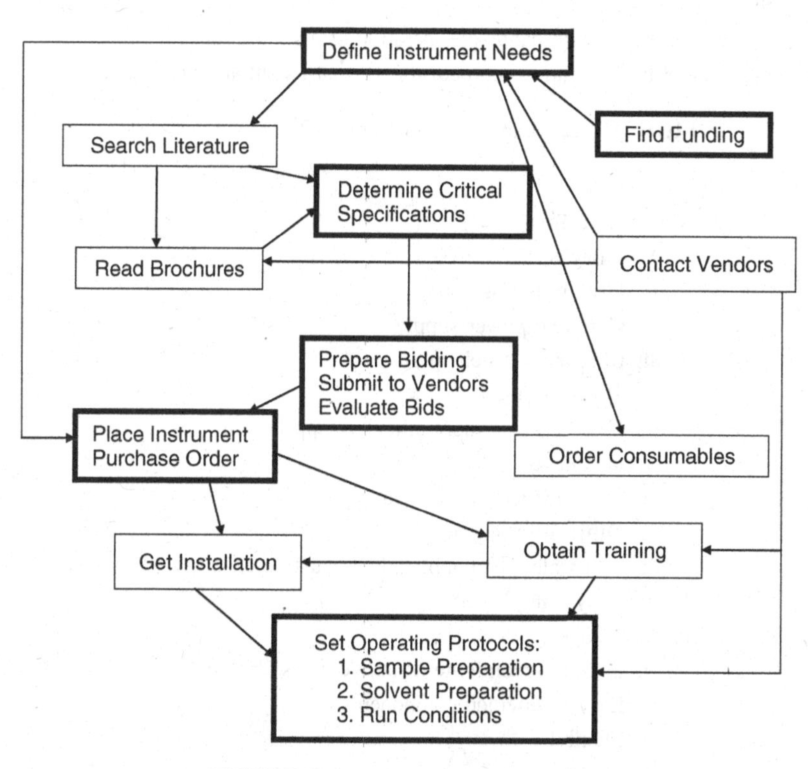

FIGURE B.1 A Purchasing Schematic

Buying and Selling Laboratory Instruments: A Practical Consulting Guide.
By Marvin C. McMaster

- Lock in needed application requirements and protocol constraints.
- View the technical literature and select required instrument(s).
- Find a knowledgeable local guru, consultant, and technical sales representative.
- Gather and evaluate sales literature to determine the specifications.
- Consider service, installation, training, and support in determining specifications.
- Consider other applications for the instrument (flexibility).
- Estimate the cost of the instrument and of required specifications.
- Find budgeting sources for additional funding, if needed.
- Build lockout specifications for the bidding proposal.
- Put out for competitive bids from a variety of vendors.
- Select bid(s) that fits your requirements.
- Write a justification letter for the desired system.
- Place the order for the selected system.
- Check instrument-shipping boxes for obvious damage. Report damage immediately.
- Arrange installation with hands-on operation of a simple run.
- Write a simplified operating protocol for the instrument and post it.

FIGURE B.2 Laboratory Instrument Purchase Checklist

1. System-in-a-Box
   a. Advantages
      i. Usually is less expensive
      ii. Common power supply
      iii. Conserves bench space
      iv. Turnkey purchases
      v. Usually has a more attractive appearance
   b. Disadvantages
      i. Often difficult to service
      ii. Difficult to upgrade
      iii. Inflexible when applications must be changed
2. Modular System
   a. Advantages
      i. Easy to reconfigure as applications change
      ii. Easy to upgrade components
      iii. Flexible
   b. Disadvantages
      i. Generally more expensive
      ii. Bench space sprawl

FIGURE B.3 System Comparison

1. Prepare 200 mL of 50% methanol/water and 100 mL of 80% methanol/water. Vacuum filter through 0.54 μm filters.
2. Remove the C_{18} column and set it aside. Set up the HPLC system with a column blank in place of the column. Prime the pump(s) with 50% methanol/water. Set over the pressure setting on the pump to 4000 psi. Set flow at 0.1 mL/min and slowly increase to 1 mL/min.
3. When the pressure is steady, turn the injector handle to inject (or load if it was already in the inject position) and watch the pressure. Cycle the injector handle to the inject position.
4. Watch the recorder or computer baseline. When it is stable, slow the pump flow to 0.1 mL/min, remove the column blank, and connect the C_{18} column to the injector. (*Do not* connect the column to the detector yet.) Wash the column solvent into a beaker (start a slow flow increase from 0.1 to 1.0 mL/min) for six column volumes (12–18 mL). Pressure should slowly increase to around 2000 psi at 1 mL/min due to column backpressure.
5. When the pressure is stable, *record column backpressure* from the pump pressure gauge in a logbook. Connect the column to the detector inlet fitting. Turn on the detector (select 245 nm, 1.0 AUFS) and the recorder at 0.5 cm/min chart speed. Observe the baseline. Drifting indicates that the detector is still warming up or something is washing off the column.
6. When the baseline is stable, inject 15 μL of column standards. Turn the injection handle quickly. Remove the injection syringe and flush three times with solvent.
7. On the chromatogram paper, mark the inject point. Record the date, time, operator name(s), flow rate, mobile phase, sample type, number, and injection amount, column, detector wavelength, and attenuation, and the chart speed so that you can duplicate this run. Record the chromatogram until the baseline is reached after the four peaks.

FIGURE B.4 Original Isocratic HPLC Run Protocol

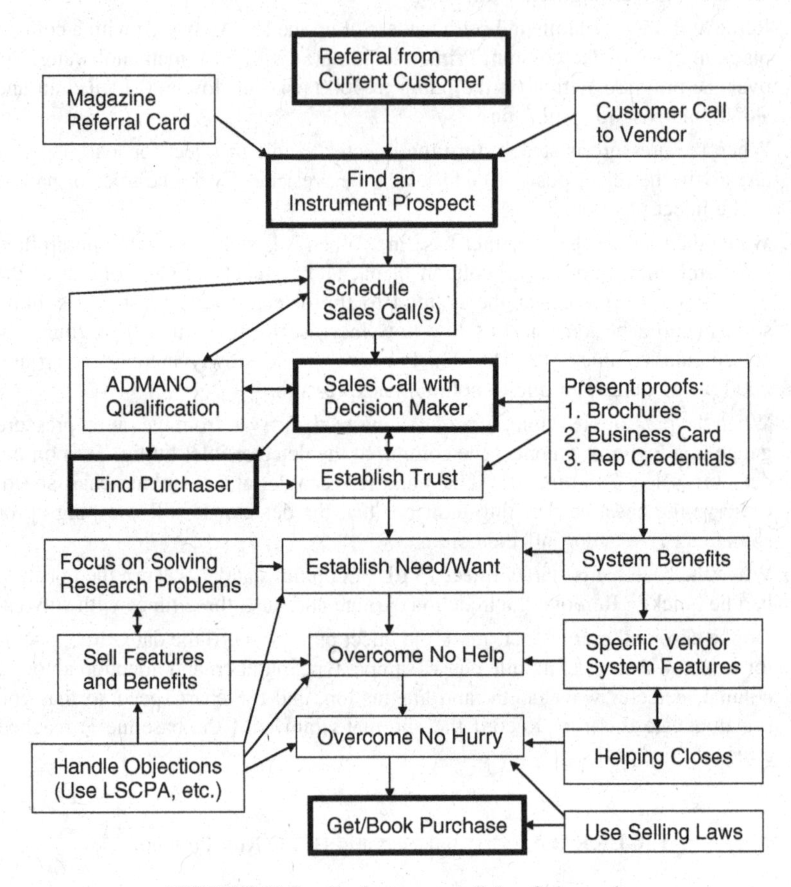

FIGURE B.5 An Instrument Sales Schematic

Application	What is the laboratory trying to do?
Decision Maker	Who approves the spending of the money?
Money	How much money is available?
Available	When will the money be available?
Next Step	What is the next step in the purchase decision?
Others	Who else is involved in the decision?

FIGURE B.6 The ADMANO Sales Interview

1. Does the prospect ask or tell (Verbal), show or mask emotions (Facial)?
2. Based on the answer to question 1, classiby the prospect:
   a. Driver—tells and masks emotions
   b. Expressive—tells and show emotions
   c. Analytical—asks and shows emotions
   d. Analytical—asks and masks emotions
3. Find the prospect's hot button from his or her category:
   a. Driver—controls money, things, people, time
   b. Expressive—wants recognition, the top position (tell this person stories)
   c. Amiable—wants to trust you, have security, use automation
   d. Analytical—wants peer respect, status, to be right, details
4. Remember to follow up with Expressives and Analyticals.

FIGURE B.7 Using HBA

1. Anyone can learn to sell.
2. You have only one chance to make a good first impression.
3. Ask questions instead of making statements.
4. Fear of loss is more important than desire for gain.
5. The answer is always "no" if you do not ask.
6. Listen to the answers when you ask questions.
7. Objections are always a sign of interest.
8. Ask for clarification instead of arguing. The phrase "help you" is wonderful to hear if it is true.
9. Learn to read and use body language to defuse tension.
10. People buy emotionally and justify the buying decision with logic.
11. People want to be fair.
12. Honesty is always good business.
13. Never criticize the competition.
14. Everything has a price. There is no such thing as a free lunch.
15. You can explain quality once or apologize for the purchase price forever.
16. Selling is always about serving.

FIGURE B.8 The Laws of Selling

TABLE B.1 Analytical Instrument Specifications

Instrument Types	Important Specifications
1. *Heater, evaporators*	Temperature range, heating rate
2. *Refrigerators, freezers*	Temperature range, cooling rate
3, *Fume hoods, bell jars*	Volume, capacity, evacuation rate
4, *Scales, pipetters*	Capacity, precision
5. *Mixers, stirrers*	Capacity, speed, torque
6. *Centrifuges*	Capacity, speed, maximum *g*-force
7. *Spectrophotometers*	Capacity, speed, detection range
(UV, IR, NMR)	+ sample specificity and sensitivity
8. *Scintillation counters*	Capacity, precision, isotope type
9. *Sequencers*	Capacity, chain-length limits
(protein, peptide, DNA)	+ step-cleavage completeness
10. *Separation systems*	Capacity, media, mobile phase
(SPE, open column, GC)	+ flow rate, detector, wavelength
(TLC, EC, 2-D EC, SCF)	+ volt range, visualizing agent
11. *Chromatography systems*	Capacity, speed. Resolution
(GC system)	+ gas, flow rate, oven temperature,
	+ temperature range and ramping,
	+ detector range and sensitivity
(SFC system)	+ SC pressure, flow rate,
	+ temperature range, detector
	+ range, and sensitivity
(HPLC)	+ mobile phase, gradient, flow,
	+ detector range and sensitivity
(GC/MS, LC/MS)	++ mass range, EM potential, and lens settings
(SFC/MS, EC/MS)	++ SC pressure and voltage

TABLE B.2 Characteristics of Buffers

Buffer (Nonvolatile)	**pKa(s)**	**Buffer Range**
1. *Phosphate—pK1*	2.1	1.1–3.1
—pK2	7.2	6.2–8.2
—pK3	12.3	11.3–13.3
2. *Borate*	9.2	8.2–10.2
Buffer (Volatile)	**pKa(s)**	**Buffer Range**
3. *Trifluoroacetic acid*	0.5	3.8–5.8
4. *Ammonium formate—pK1*	3.8	2.8–4.8
5. *Ammonium acetate—pK1*	4.8	3.8–5.8
6. *4-Methylmorpholine*	8.4	7.4–9.4
6. *Ammonium bicarbonate*	6.3/9.2/10.3	6.8–11.3
7. *Ammonium acetate—pK2*	9.2	8.2–10.2
8. *Ammonium formate—pK2*	9.2	8.2–10.2
9. *1-Methylpiperidine*	10.1	10.0–12.0
10. *Triethylammonium acetate*	11.0	10.0–12.0
11. *Pyrrolidine*	11.3	10.3–12.3

TABLE B.3 Characteristics of Common Solvents

Solvent	Formula	MW (daltons)	Boiling Pt. (°C)	UV Cutoff (nm)
Acetonitrile	CH_3CN	41.05	81.6	190
Chloroform	$CHCl_3$	119.38	61.7	245
Dichloromethane	CH_2Cl_2	84.93	40.0	235
Ethanol	CH_3CH_2OH	46.08	78.5	210
Ethyl acetate	$CH_3CO_2CH_2CH_3$	88.12	77.1	260
Diethyl ether	$(CH_3CH_2)_2O$	74.12	34.5	220
Heptane	$CH_3(CH_2)_5CH_3$	100.21	98.4	200
Hexane	$CH_3(CH_2)_4CH_3$	86.18	69	200
Methanol	CH_3OH	32.04	65	205
n-Propanol	$CH_3CH_2CH_2OH$	60.11	97.4	210
Isopropanol	$CH_3CH(OH)CH_3$	60.11	82.4	210
Tetrahydrofuran	C_4H_8O	72.12	66	215
Toluene	$C_6H_5(CH_3)$	92.15	110.6	285
Water	H_2O	18.02	100	none

APPENDIX C

GLOSSARY OF PURCHASING AND SALES TERMS

ADMANO sales interview A systematic client interview procedure designed to provide the application, personnel, and financial information necessary to position a sale.

Amiable A personality type motivated by security and trust. These persons tend to ask questions and show emotions. They want to trust you and be your friend. They want security and like automated equipment.

Analytical A personality type motivated by status or peer respect. These persons tend to ask questions and mask emotions. They thrive on details.

Application laboratory Vendor or private contractor that carries out methods and application research as a customer service or on a fee basis.

Automation Instrument add-on that equips a system for continuous, unattended operation. In modern usage, automation usually is computer controlled.

Benefit An advantage to a customer of one of the features of the instrument or system being addressed by the sales presentation.

Buying and Selling Laboratory Instruments: A Practical Consulting Guide.
By Marvin C. McMaster

Bidding An instrument specifications list submitted to vendors in order to obtain competitive pricing. Usually used in university or governmental accounts where purchasing will not be made from GSA price lists.

Body language A form of nonverbal communication that uses the positioning of the body. About 70% of all communication is estimated to be non-verbal. Agreement between the words said, the facial expression, and the body language is extremely helpful in establishing the truth.

Closing The ending of a sales presentation; an attempt to achieve a customer's agreement and obtain a purchase agreement.

Consumables Items necessary to the operation of the instrument that are used up and must be replaced to continue operation.

Data archival Backed-up and stored information produced by the instrument; this can be in the form of raw data or reports.

Demonstration equipment Equipment used to prove the effectiveness of an instrument or a system for a particular customer application; often offered at a discount price to achieve a sale.

Desire and need Enthusiasm generated for the instrument based on perceived value by the customer.

Discounting A form of sales incentive for the customer that consists of lowering the price in some manner; discounts can consist of 10% off the price of demonstration equipment, supplies, training, or support. Quantity discounts are offered to obtain multi-instrument purchases.

Driver A personality type motivated by control of people, time, or money. Drivers tend to tell and mask emotions; they want to control buying decisions, people, time, money, and things.

Expressive A personality type motivated by recognition. These persons tend to tell and show emotions. They want to be number one. Tell them stories; follow up on the sale until the instrument is up and running in their laboratory.

Fear of loss Avoiding loss is more important in the decision process that the desire for gain. Khrushchev said, "What's mine is mine. What's yours is negotiable."

Feature A characteristic of the instrument that may interest and be valuable to the customer. Each feature needs to be turned into a customer benefit to be of value in a sales presentation.

Grant Financial support from a funding agency. A major function of the laboratory director is to write grant proposals so that the work of the laboratory can proceed. Grants provide needed instruments, consumables, and salaries.

GSA pricing The Government Service Administration (GSA) negotiates the best prices with manufacturers and makes these prices available in a printed price list to government laboratories for instrument purchases. The government is the single largest purchaser of instrument in the United States, and expects and insists on the lowest prices. Violators can be removed from the GSA price list, which greatly reduces their profitability.

Hot button analysis (HBA) A technique for quickly determining a person's personality type, the person's major motivator, and a possible way of selling to that "hot button"; a way of tailoring the sales presentation to the customer's major interest and handling problems when they arise.

Justification letter A letter sent with returned bid proposals by an investigator providing reasons for purchasing a system that was not the lowest competitive bid based on failure of the lowest bid to meet bidding specifications.

Listening A salesperson's best friend. Questions and careful, attentive listening are the major sources of information needed to complete a sale. When a salesperson is talking, both the salesperson and the customer are losing.

Lockout specifications Specifications of the instrument that favor the desired instrument over other possible alternatives. These specifications are used in competitive bidding to allow the customer to obtain exactly the equipment needed to do the work in the laboratory, yet allow flexibility for innovation.

Manuals Complete document on instrument components, design, operation, troubleshooting, and parts originally supplied with all instruments; recently, only limited documentation has been included, with full documentation available on disk in help screens, online on web sites, from the manufacturer on request. Sometimes no manual exists.

Manufacturers The companies that make the instrument under consideration. Many manufacturers make only part of the instrument and include equipment from other equipment manufacturers (OEMs). Usually the manufacturer selling the instrument takes responsibility for the quality of the whole instrument. Vendors may be manufacturers but may also sell equipment made by others.

Modular system Component system integrated with cables, software, and tubing; modules are usually stackable and replaceable, allowing flexibility for rapid upgrading with newer components.

Objections A sign of customer interest. An apparently negative comment about the instrument made by a prospect, that when answered and handled, can become a major reason for the prospect to buy.

Purchasing The process of agreeing to buy an instrument at a given price. A purchasing agreement is the document announcing the decision to buy. The Purchasing Department is a customer's organization involved in submitting bids to vendors, helping the customer to evaluate bid proposals from vendors, and issuing purchase orders when the purchaser makes a decision.

Purchasing agent Employee in the Purchasing Department with responsibility for submitting bids to vendors, checking received bids for adherence to bidding specifications, helping purchasers to make a decision, and issuing purchase orders.

Questions Verbal tools for gathering information from a prospect to aid in making a sale. There are two types of questions: *open questions* are "who, where, why, what, how" questions used in unstructured gathering of information. *Closed questions* guide prospects to "yes" and "no" answers to move the selling process toward completion.

Recycling and disposal Process of moving an instrument out of the laboratory or into other applications when it is no longer useful for the application for which it was purchased. If it is moved out of the laboratory, it may be donated, resold, or scrapped.

Reverse order diagnosis A method of systematically solving a system problem to determine the module or part of a module causing the problem. It usually starts with the data acquisition module, validating its performance, then using the data system to work up the system toward the sampling end and validating each component in its turn until the culprit is found.

Sales justification The evidence provided by the sales representative to convince the prospect to buy; this includes the material that created the emotional decision to buy (desire) and the logical features that support a logical (need) justification for the purchase.

Sales representative The local representative of a vendor who has responsibility for making the instrument sale. He or she can be a professional salesperson (win/win) who acts as a consultant to the laboratory, ensuring that the correct system is purchased, or a con artist (win/lose) who tries to sell only the instrument the salesperson's company wants sold, regardless of the customer's needs.

Sales tools The materials used by the sales representative that make the customer trust the salesperson and his or her company, realize the need and desire for the equipment, and convince the customer that this equipment will solve his or her problem and must be purchased now. Tools include the salesperson's business card, company and instrument brochures, technical articles using the company's equipment, the sales presentation, and referrals to successful customers.

Sales urgency Customers are often hesitant to release the money to make a purchase. They make few major decisions a year and are not familiar with the process, they worked hard to acquire the funds now at their disposal, and the laboratory has other needs for the money. Urgency must be created by handling objections, evoking desire, demonstrating value, and asking for the order.

Sample preparation Actual samples are heavily contaminated. Simplification by rapid purification can speed analysis and increase the sensitivity of detection. Filtration, extraction, and SFE purification can increase analysis speed and sensitivity.

Seminars Professional scientific presentations by company representatives on the use of an instrument can prove its capability for specific analysis and aid the selling process.

Service representative The availability of a local professional field service technician supported by a manufacturer's service department can be critical to ensure the customer's success and continued instrument operation.

Specifications Detailed features of the instrument that can be turned into benefits for customers to apply to their application needs. Not all specifications can be turned into benefits; the critical specifications need to be isolated to be used in competitive bidding.

System-in-a-box An organized instrument system, often with a common power supply for all modules, that is sold for a specific purpose. It is usually less expensive than other systems and therefore is attractive. It is often difficult to service and upgrade.

TANSTAAFL A fact of life: there ain't no such thing as a free lunch. The desire to gain something for nothing is deeply ingrained but is unrealistic. Everything has a price, although this is not always immediately apparent.

Training Customer schools designed to improve the operation of the instrument. This can take the form of a seminar, hands-on training at the customer's location, or training at the manufacturer's facility—a valuable sales incentive to be offered by the salesperson.

University instrument donation A method of disposing of obsolete equipment no longer needed by the laboratory by giving the system to a local educational institution for a tax deduction (win/win).

Vendor The company that sells the instrument. This may or may not be the original manufacturer, but it is the entity that offers the instrument for purchase and has ultimate responsibility for the instrument's successful performance.

Warranty Guarantee offered by the manufacturer or vendor that the instrument will work as advertised. It is usually only as good as the company making the promise.

Win/lose sale A sales situation in which one party, usually the salesperson, benefits more than the purchaser. Sometimes the equipment sold is more elaborate and expensive than that needed to achieve the buyer's needs; sometime it just does not work.

Win/win sale A situation in which both parties benefit from the sale. When the sale is complete and the instrument is installed, it quickly begins to generate results for the customer and continues to operate with little downtime. The company makes a profit and the salesperson receives a salary, a bonus, a compliment, and a good referral from a successful customer.

APPENDIX D

TROUBLESHOOTING QUICK REFERENCE

This section is designed for troubleshooting problems associated with purchasing and selling laboratory instruments. Following is a series of commonly encountered purchasing problems, possible causes, and suggested solutions.

D.1 TROUBLESHOOTING THE PURCHASE

Problem 1.1: Prioritize Among Needed Pieces of Laboratory Equipment

Solution a: List research applications and needed equipment estimates and pricing.

Solution b: Determine immediate funding sources and amounts.

Solution c: Prioritize the importance and immediacy of each project.

Problem 1.2: Know What Research Equipment Is Needed

Solution a: Read technical literature on the application. Determine the type of equipment used for the analysis or separation.

Buying and Selling Laboratory Instruments: A Practical Consulting Guide.
By Marvin C. McMaster

Solution b: Visit major technical meetings with instrument exhibits. Talk to various vendors about needed equipment.

Solution c: Talk/phone colleagues doing the required type of purification or analysis.

Solution d: Reply to interest cards from magazines offering information on desired instruments.

Problem 1.3: Determine Equipment Pricing and Specifications

Solution a: Call or access the vendor's web site and request literature on desired equipment.

Solution b: Call the company and request a visit from the local sales representative. Try to find a knowledgeable, helpful contact.

Solution c: Request a price quotation for the best-fit instrumentation.

Solution d: Write grant proposals based on the literature and on specifications for addition funding for the project. Consider both government and private granting sources.

Problem 1.4: Select Vendors for Needed Equipment

Solution a: Match brochure specifications against the application's needs.

Solution b: Talk to vendors' sales representatives. Select representatives who understand the application and the equipment.

Solution c: Look for the best fit between the application and the instrument's features, and for upgrade flexibility, good service, and a good reputation for support.

Problem 1.5: Select Specifications for Bidding

Solution a: Find specifications and features needed for the application among all the brochures and vendors. Build a list of desired specifications for a bidding proposal.

Solution b: Write lockout bid specifications specific to the instrument and the representative that best fits your needs. (Remember, companies and sales representative change and disappear.)

Solution c: State that instrument manuals with hardware diagrams and parts lists must be included in the bidding as part of the order.

Solution d: Require included training and support to be spelled out as part of the bidding proposal.

Problem 1.6: Determine Needed Consumables and Costs

Solution a: Contact referrals from sales representatives for the approximate cost of supplies and consumables per year.

Solution b: Contact manufacturers for their estimates on the cost of consumables.

Solution c: Contact local gurus and instrument users to verify their consumable costs.

Solution d: Talk to users at technical meetings about their estimated cost of consumables.

Problem 1.7: Get the Best System from the Bid Results

Solution a: Review bids and reject those that are too expensive and that do not meet important bidding specifications.

Solution b: Write a justification letter to the Purchasing Department stating the reasons for eliminating bids that are competitively priced but do not meet the laboratory's needs. Use lack of support, training, and manuals as part of the justification and specifications misfit for the rest.

Solution c: If none of the bids match your needs, reject all of them and rebid with modified bidding specifications.

Problem 1.8: Place the Order for Fast Delivery

Solution a: Get bid proposals back to the Purchasing Department along with your justification letter for the selected bid as soon as possible.

Solution b: Talk to the purchasing agent about quickly issuing the purchase order and get permission to express-mail it to the vendor.

Solution c: Talk to the sales representative about quickly booking the order, getting it into the manufacturing queue, and expediting shipping.

Solution d: Ask the sale representative to alert the field service representative about the tentative ship date in order to book the system installation as early as possible.

Problem 1.9: Obtain Training for Operators

Solution a: Schedule operators for the manufacturer's training school as soon as the purchase order is booked.

Solution b: Talk to the sales representative about providing hands-on training and startup assistance as soon as the instrument is installed.

Solution c: Talk to the service technician about service tricks, expected problems, and a list of needed tools, fittings, and spare parts.

Solution d: Make sure that you get the system manuals, not just Read-Me-First brochures.

Problem 1.10: Find Service When Problems Occur

Solution a: Get the service technician's business card with his name and voicemail number. Post a copy on the laboratory bulletin board and on your secretary's Rolodex. Add the salesperson's name and voicemail number.

Solution b: Ask your salesperson for the name and telephone number of the in-house person who knows where everything is kept and everyone who can help when things go wrong.

Solution c; Get the telephone numbers of the field service manager and the in-house service manager. Call, introduce yourself, and tell them how well the installation went.

Solution d: Find someone in the university repair shop who is not afraid of hardware and electronics. Show him the system manuals and ask if they make sense. Help him with his tool budget. Make a friend in the facility who can be there for an emergency first response.

Solution e: Find a third-party instrument service supplier who has experienced service technicians formerly associated with vendors.

D.2 TROUBLESHOOTING THE SALE

Problem 2.1: Find Prospects Who Can Buy

Solution a: Get referrals to potential prospects from current users of your company's equipment.

Solution b: Attend a local American Chemical Society (ACS) meeting and meet users of other companies' equipment. Advise them about how to solve problems on which their vendors are not helping. Get referrals to potential customers.

Solution c: Follow up as soon as possible on lead lists from magazine advertising cards. Call and schedule an appointment with the principal investigator (PI).

Solution d: Visit companies that should be using your type of equipment and ask employees who runs their laboratory.

Solution e: Do booth duty at a technical meeting, talk to visitors, and get the names of investigators they know in your territory who might be using or could use your equipment.

Problem 2.2: Find the Decision Maker, Application, and Money

Solution a: Use ADMANO qualifying questions with investigators and laboratory personnel to find an application and a decision maker who can use your equipment. Use the contact to introduce yourself to the decision maker and make a sales appointment. Get information on money, availability, the next step, and other people who influence the sale.

Solution b: Three-second sales appointments are best; find out if this is a good time to discuss your instrument and the customer's application.

Second c: Do an immediate HBA on the decision maker and adjust your sales approach to match his or her style.

Problem 2.3: Make the First Sales Call and Establish Trust

Solution a: Give the prospect your business card; mention where you got his or her name and why you think you may be able to help with the application. Introduce yourself and your company and use consultant selling.

Solution b: Ask the prospect if he or she knows other people doing similar separations. Get the prospect a list of research papers that mention the use of your equipment in doing similar separations.

Solution c: Offer the prospect a list of laboratories that are using your equipment successfully.

Problem 2.4: Create Desire and Prove That the Salesperson's Company Can Help

Solution a: Use open questions (why, who, what, where, how) to find out exactly what the prospect needs to do and how he or she is doing it now.

Solution b: Finish the ADMANO interview to determine how the prospect will fund the instrument purchase, when, and what other people must approve or help with the instrument purchase. Find out who will use the instrument in the laboratory.

Solution c: Give the prospect brochures on the specific instrument and explain how it has been used on similar problems. Show how your instrument can save time, solve the prospect's research problems, and help the prospect write papers and train graduate students.

Solution d: If you have had personal experience with the instrument in solving problems, discuss your experience and success. Offer your help in getting the prospect's system up and running.

Solution e: Discuss the company's training school and support to guarantee the prospect's success. If the company does not offer them, put them together at your location so that you can offer them.

Solution f: Use a closed question (which requires a Yes/No answer) to ensure that the prospect understands the power of the instrument: "Let me show you how I think our system can make a difference in your research. Would that be all right?"

Problem 2.5: Get, Book, and Track the Order

Solution a: Use LSCPA to ask for the order or for the bid on your locked-in specifications. Listen to any objections and use "feel, felt, found" to handle them, clarify any misunderstanding, propose that the prospect request a purchase order (PO) or help him to prepare lockout specifications for bidding, and ask him to send it to the Purchasing Department.

Solution b: If the prospect gives you a PO, fax or express mail it to the company, track it with Order Entry to ensure that it is booked, and get it into the system for shipping.

Solution c: If the prospect is bidding, get permission to walk the bid proposal to the Purchasing Department. Track the posted bid to see who the competition will be and what the closing date is. Once you have won the bid, get a tentative shipping date and book the installation.

Solution d: Visit the prospect, give the closing date, and ask if he or she will need help writing the justification letter for your system (assumptive close).

Solution e: If allowed, walk the bidding decision and justification letter to the Purchasing Department, and ask when the PO will be issued and if you can mail it in for the prospect. Use the idea that you are trying to provide a service. Be careful! Sometimes this approach seems to be pushy. "Help you" are magic words, but they can be overused.

Problem 2.6: Contact the Laboratory and Arrange Training

Solution a: Call or visit your laboratory contact and arrange to bring in the laboratory's operators for company training. The actual decision maker is probably someone in the laboratory; this person needs to be kept informed about shipping, installation, and training.

Solution b: If no company training is available and the instrument can be used, offer to help train the operators after installation.

Solution c: If you cannot provide training, arrange help from an in-house specialist or ask the company to send in a laboratory specialist. Arrange an informal seminar at the customer's location and combine this with on-site training.

Problem 2.7: Arrange Installation and First-Time Use

Solution a: Call the service technician and schedule a tentative installation as soon as the shipping date is set.

Solution b: Call the customer's laboratory to give the shipping date. Have their personnel check for external damage on shipping boxes on arrival and contact the company and the insurer if any damage is found.

Solution c: Make sure that the customer has operating supplies and standards to check the instrument's operation on the first run.

Solution d: Prepare a typed first-run protocol, make an appointment after installation to post a copy of the protocol, help the operator go through the first run with standards, and leave the protocol to ensure that the client will continue to use the instrument.

Problem 2.8: Get Referrals and Follow-up Sales

Solution a: When the instrument is installed, up and running, and being used successful on the application, ask the PI if you can use the customer's name as a successful referral to other investigators.

Solution b: Ask the PI and the laboratory decision maker who they know in the university or at other accounts who might be buying similar equipment.

Solution c: If you get referral, ask if you can use the customer's name when using it.

Solution d: Ask the referrer if he or she would like to brag about the new instrument to the referral and introduce you over the phone or by e-mail if the referrer's account is in your territory.

APPENDIX E

SELECTED READING LIST

Bettger, Frank. *How I Raised Myself from Failure to Success in Selling*. Prentice-Hall, New York, 1949.

Carnegie, Dale. *How to Win Friends and Influence People*. Simon and Schuster, New York, 1936.

Girard, Joe. *How to Sell Anything to Anybody*. Warner Books, New York, 1977.

Lorayne, Harry and Lucas, Jerry. *The Memory Book*. Stein and Day, New York, 1974.

Mandino, Og. *The Greatest Salesman in the World*. Fredrick Fells, New York, 1975.

McMaster, M. C. *LC/MS: A Practical User's Guide*. John Wiley & Sons, Hoboken, NJ, 2005.

McMaster, M. C. *HPLC: A Practical User's Guide*, 2nd ed., John Wiley & Sons, Hoboken, NJ, 2007.

McMaster, M. C. *GC/MS: A Practical User's Guide*, 2nd ed., John Wiley & Sons, Hoboken, NJ, 2008.

Schwartz, David J. *The Magic of Thinking Big*. Simon and Schuster, New York, 1959.

Buying and Selling Laboratory Instruments: A Practical Consulting Guide.
By Marvin C. McMaster

Watson, Tom. *How to Master the Art of Selling*. Champion Press, Scottsdale, AZ, 1982.

Ziglar, Zig. *See You at the Top*. Ziglar Corporation, Dallas, TX, 1975.

Ziglar, Zig. *Secrets of Closing the Sale*. Fleming H. Revell, Old Tappan, NJ, 1984.

INDEX

Buying and Selling Laboratory Instruments: A Practical Consulting Guide.
By Marvin C. McMaster